THE
CEO OF
TECHNOLOGY

Founded in 1807, John Wiley & Sons is the oldest independent publishing company in the United States. With offices in North America, Europe, Asia, and Australia, Wiley is globally committed to developing and marketing print and electronic products and services for our customers' professional and personal knowledge and understanding.

The Wiley CIO series provides information, tools, and insights to IT executives and managers. The products in this series cover a wide range of topics that supply strategic and implementation guidance on the latest technology trends, leadership, and emerging best practices.

Titles in the Wiley CIO series include:

The Agile Architecture Revolution: How Cloud Computing, REST-Based SOA, and Mobile Computing Are Changing Enterprise IT by Jason Bloomberg

Architecting the Cloud: Design Decisions for Cloud Computing Service Models by Michael J. Kavis

Big Data, Big Analytics: Emerging Business Intelligence and Analytic Trends for Today's Businesses by Michael Minelli, Michele Chambers, and Ambiga Dhiraj

The Big Shift in IT Leadership: How Great CIOs Leverage the Power of Technology for Strategic Business Growth in the Customer-Centric Economy by Hunter Muller

The CEO of Technology: Lead, Re-imagine and Invent to Drive Growth and Value in Unprecedented Times by Hunter Muller

The Chief Information Officer's Body of Knowledge: People, Process, and Technology by Dean Lane

CIO Best Practices: Enabling Strategic Value with Information Technology (Second Edition) by Joe Stenzel, Randy Betancourt, Gary Cokins, Alyssa Farrell, Bill Flemming, Michael H. Hugos, Jonathan Hujsak, and Karl Schubert

The CIO Playbook: Strategies and Best Practices for IT Leaders to Deliver Value by Nicholas R. Colisto

Cloud Computing and Electronic Discovery by James P. Martin and Harry Cendrowski

The Complete Software Project Manager: Mastering Technology from Planning to Launch and Beyond by Anna P. Murray

Confessions of a Successful CIO: How the Best CIOs Tackle Their Toughest Business Challenges by Dan Roberts and Brian P. Watson

Enterprise Performance Management Done Right: An Operating System for Your Organization by Ron Dimon

Executive's Guide to Virtual Worlds: How Avatars Are Transforming Your Business and Your Brand by Lonnie Benson

Information Governance: Concepts, Strategies, and Best Practices by Robert F. Smallwood

IT Leadership Manual: Roadmap to Becoming a Trusted Business Partner by Alan R. Guibord

Leading the Epic Revolution: How CIOs Drive Innovation and Create Value Across the Enterprise by Hunter Muller

Managing Electronic Records: Methods, Best Practices, and Technologies by Robert F. Smallwood

On Top of the Cloud: How CIOs Leverage New Technologies to Drive Change and Build Value Across the Enterprise by Hunter Muller

Straight to the Top: CIO Leadership in a Mobile, Social, and Cloud-based World (Second Edition) by Gregory S. Smith

Strategic IT: Best Practices for Managers and Executives by Arthur M. Langer and Lyle Yorks

Transforming IT Culture: How to Use Social Intelligence, Human Factors, and Collaboration to Create an IT Department That Outperforms by Frank Wander

Unleashing the Power of IT: Bringing People, Business, and Technology Together (Second Edition) by Dan Roberts

The U.S. Technology Skills Gap: What Every Technology Executive Must Know to Save America's Future by Gary J. Beach

THE
CEO OF
TECHNOLOGY

**Lead, Reimagine, and Reinvent to Drive
Growth and Create Value in Unprecedented
Times**

HUNTER MULLER

WILEY

Published by John Wiley & Sons, Inc., Hoboken, New Jersey.
Published simultaneously in Canada.

For general information on our other products and services or for technical support, please contact our Customer Care Department within the United States at (800) 762-2974, outside the United States at (317) 572-3993, or fax (317) 572-4002.

Wiley publishes in a variety of print and electronic formats and by print-on-demand. Some material included with standard print versions of this book may not be included in e-books or in print-on-demand. If this book refers to media such as a CD or DVD that is not included in the version you purchased, you may download this material at http://booksupport.wiley.com. For more information about Wiley products, visit www.wiley.com.

Library of Congress Cataloging-in-Publication Data

Names: Muller, Hunter, author.
Title: The CEO of technology : lead, reimagine, and reinvent to drive growth and create value in unprecedented times / by Hunter Muller.
Description: Hoboken, New Jersey : John Wiley & Sons, Inc., [2018] | Series: Wiley CIO | Includes index. |
Identifiers: LCCN 2017046172 (print) | LCCN 2017052892 (ebook) | ISBN 9781119276319 (pdf) | ISBN 9781119276357 (epub) | ISBN 9781119270232 (cloth)
Subjects: LCSH: Information technology—Management. | Technological innovations—Management. | Chief information officers. | Chief executive officers. | Leadership.
Classification: LCC HD30.2 (ebook) | LCC HD30.2 .M849464 2018 (print) | DDC 658.4/2—dc23
LC record available at https://lccn.loc.gov/2017046172

Cover Design: Wiley
Cover Image: © ChrisHepburn/iStockphoto

Printed in the United States of America

10 9 8 7 6 5 4 3 2 1

For Brice and Chase

CONTENTS

PREFACE

CREATING A CULTURE OF GENIUS: LEAD, REIMAGINE AND REINVENT

I've spoken often about the challenges facing Apple in markets where functionality is more highly valued than aesthetics. I've also written about how the spirit of Steve Jobs lives on in every Apple product. When you pick up an Apple device, you feel a magical connection with a legendary genius.

That's why I'm not too worried about the future of Apple. Yes, Apple will face increasingly harsh competition from other tech giants such as Google, Facebook, Microsoft, and Amazon. But Apple will survive, and here's why: Apple has a culture of genius.

The genius culture knows no limits. The genius culture dreams the impossible. The genius culture sets the imagination free and lets it run wild. The genius culture has no boundaries. It is truly global and universal. It's a leader, not a follower.

I sincerely believe in its long-term value as a foundational element in strategic leadership and business growth.

The culture of genius unlocks unlimited potential, not only within the organization, but also beyond its boundaries. It is a culture that drives and enables continuous improvement and reinvention. It is a platform for creative disruption and relentless innovation in core, adjacent, and new markets. Essentially, it is a culture that radically expands potential and continually opens doors to growth.

It is also a learning culture, in the sense that you're never relying totally on your own abilities and knowledge. Establishing and sustaining a culture of genius requires a steady intake of new information. You are always in the learning mode, always looking for new ideas and new trends that can be turned rapidly into new products and services.

The culture of genius is a virtuous circle, fed by learning, analyzing, developing, and testing—all at extremely high speed. The goal is getting new products and services to market before your competitors even realize there's a need for those products!

Continuous innovation requires incredible levels of discipline, strength, and agility. From my perspective, those qualities are now table stakes. If you don't have them, you're not really in the game. But you need a culture of genius that consistently supports, encourages, embeds, and rewards those qualities over the long haul. It can't be a two-week program; it needs to become the deep culture of the organization.

Genius as a Core Capability

What do companies like Apple, Amazon, Facebook, and Tesla have in common? The answer is simple: They have a culture of genius.

The culture of those companies is their core strength. It's the foundation upon which they build and achieve incredible success, year after year. Cultures of genius don't sit still; they lead, innovate, reimagine, and reinvent the world around them.

Make no mistake: A genius culture doesn't just happen. It's not an overnight phenomenon. It takes years of courage, discipline, hard work, and effort to build a genius culture and to grow it successfully. It requires visionary leadership and deep understanding of how modern markets work.

It also takes nerves of steel. Building a genius culture isn't a part-time job. It requires your attention, your presence, and your commitment.

I sincerely believe that the genius culture concept is both real and valuable. I invite you to join our quest to learn more about the culture of genius, and to join our explorations as we look for the core traits of genius in the twenty-first century.

I'll take it a step further: Organizations that build and nurture cultures of genius will thrive and succeed; organizations that stick with the status quo will wither and die.

The value proposition is clear: Companies with a culture of genius will perform at levels that are far above their rivals. They will leave their competitors in the dust, wondering what happened.

Having a culture of genius enables your organization to recognize subtle shifts in markets, respond rapidly with practical go-to-market strategies, and capture first-mover advantages before anyone else is even aware of the opportunities.

A culture of genius creates its own momentum, racing ahead of the pack and establishing front-runner status. Instead of following the herd, you inspire, uplift, and motivate your teams, your business partners, your supply chains, and your customers to continually seek the next level of greatness.

A culture of genius isn't afraid of the future—it embraces the future and revels in its unlimited potential.

ACKNOWLEDGMENTS

This is my fifth book, and like its predecessors, this book represents several years of in-depth research and analysis.

I am honored to acknowledge the contributions of our expert sources, who shared their time, insight, and experiences freely and without reservation. I personally thank Snehal Antani, Helen Arnold, Ramón Baez, Asheem Chandna, Guy Chiarello, Mignona Cote, Lee Congdon, Jason Cooper, Kirsten Davies, Dana Deasy, Scott Fenton, John Foley, Pat Gelsinger, Clark Golestani, Patty Hatter, Kevin Haskew, Shawn Henry, Zack Hicks, Adriana Karaboutis, Tim Kasbe, Ralph Loura, Shamim Mohammad, Tom Peck, Mark Polansky, John Rossman, Bernadette Rotolo, Bill Ruh, Kevin Sealy, Raj Singh, Naresh Shanker, Jim Swanson, Joe Topinka, Michael Wilson, and Deanna Wise for contributing their expertise and wisdom to this book.

I also thank my excellent team here at HMG Strategy: Kimberly Ball, Darien Bisson, Travis Drew, Tom Hoffman, Melissa Marr, Peggy Pedwano, Lindsay Prior, and Kristen Sciliano.

Additionally, I thank Mike Barlow, who served as editorial director on this book project. Mike's professional advice and guidance were invaluable.

And of course, I thank my editor, Sheck Cho, for his generous support and dedication over the many years we've worked together.

INTRODUCTION

ARE YOU READY FOR 2030?

Roughly 2 million years ago, our Stone Age ancestors discovered how to control fire. Ever since, progress and innovation have defined human culture. We invented the crossbow, printing press, steam engine, electricity, internal combustion engine, air travel, atomic power, digital information technology, and space travel. Each invention rocked our world and moved us further into an unknown future. Despite our uncertainty, we keep moving forward.

Think of the amazing visionaries who have shaped our modern world: Newton, Watt, Curie, Einstein, Edison, Ford, Gates, Jobs, and Musk. They saw the future and they remade the universe. They are the role models and the avatars of meaningful progress and transformation.

Today, the pace of innovation is accelerating at a rate that would have been simply unimaginable 10 years ago. By 2030, the world will be a completely different place than the world our parents knew.

Artificial intelligence, bioengineering, ubiquitous computing, advanced manufacturing, driverless cars, drones, nanobots, blockchain, and new financial technologies (fintech) will fundamentally and profoundly transform every aspect of our lives.

So here's my question: Are you ready for 2030? Are you making the right investments and focusing on the right strategies for a world that does not resemble anything we could have imagined in the past? Are you preparing for a future of unprecedented change and transformation, at every level of society and across all industries?

If you're not, you and your company will not survive. You will be swept away by a flood of competitors who have learned how to innovate more quickly and more effectively. You will face a deluge of competition from companies that are more nimble, more agile, and more able to capitalize on new ideas quicker than you can.

This isn't idle speculation. This is both a challenge and a warning: Innovate or die. Yes, it's difficult to predict the future. But it's not difficult to predict that the success of your enterprise will depend on your ability to prepare for future scenarios that are wildly different from anything we are experiencing today.

Are you ready for that kind of change and transformation? Do you have the infrastructure, the talent, and the resources necessary to move swiftly and take advantage of opportunities at the moment they arise? Do you have the courage, expertise, and experience necessary to create new markets for new kinds of products and services? Are you ready to lead the revolution, and not just follow the herd?

From my perspective, the greatest and most effective leaders are also the most visionary leaders. So let's begin by asking this question: What's your vision for 2030?

The Burning Platform

We live in unprecedented times. Markets are chaotic. Consumer demand is unpredictable. Product cycles have tightened drastically. Regulations, rules, and guidelines are changing rapidly. Everywhere you look, there is turbulence and upheaval. With amazing force and astonishing speed, the paradigm has shifted.

As leaders, we face continuous disruption and extremely rapid transformation. The C-suite and board of directors look to the CIO for expert guidance and flawless execution.

Ask yourself: Are you ready for your next meeting with the C-suite? Are you ready for your next meeting with the board of directors?

Do you have the passion, commitment, energy, and motivation to lead your organization through truly perilous times?

Are you genuinely interested in people? Do you understand what people need to do their best work? Are you willing to help the people around you succeed and thrive? Are you generous, collaborative, and cooperative? Do you have the stamina and skills required to lead high-performance teams?

Never before have the stakes been so high. The risks are clear: Make the wrong decision and your company will fall behind, suffering potentially irreparable damage to its brand and reputation.

Pressure to Innovate

The C-suite faces crushing pressure to create and deliver innovative business strategies for growth and success in modern competitive markets.

CIOs and chief information security officers (CISOs) confront a daunting array of challenging tasks and responsibilities. Expectations have risen dramatically: The digital customer experience must be fast, flawless, and completely secure.

Speed, Agility, and Creativity

For the twenty-first-century enterprise, success will depend on speed, agility, creativity, and excellence at all levels. But achieving phenomenal success will require more than just strong business acumen and superior leadership skills—it will require true digital fluency and an extraordinarily deep understanding of new technology.

This book provides a meticulously detailed roadmap to success in the age of digital transformation. It is written expressly for top executives, board members, investors, innovators, and entrepreneurs. It is a uniquely valuable collection of insight, experience, and first-hand knowledge, collected from the very best minds of our generation and gathered into one absolutely essential book.

I invite you to join me on a journey of discovery. This book will enhance and deepen the wisdom, experience, and insight that you already possess. It will make you a better, more focused, and more successful leader.

Critical Insights

In this book, we will dive deeply into key imperatives that define the CIO's mission in the modern enterprise, such as:

- Reimagining the digital customer experience and delivering value at every touchpoint across the enterprise
- Enabling innovation and empowering the CEO's vision for growth in core, parallel, and new markets
- Security across the enterprise and beyond
- Hiring and retaining top talent
- Simplifying IT and making it easier to use
- Building strong partnership across the C-suite and with strategic vendors
- Creating an agile culture focused on speed to cash, performance, and innovation

Call to Action

This is my fifth book on the critical subject of technology executive leadership. It introduces a game-changing and truly visionary approach to the challenges of integrating innovative technology and modern business strategy to generate value for the twenty-first-century enterprise.

It is absolutely imperative for technology executives to become exceptional leaders and great communicators of business strategy across the modern enterprise.

IT is experiencing a seismic shift of incredible magnitude. Compared to the IT transformations of the 1990s and 2000s, today's events seem truly revolutionary. The stakes are higher, everything is in flux, and there are no guaranteed paths to success.

Great CIOs, CTOs, and CISOs are great team leaders. They are like great CEOs—they understand that it's always about the people. New technology is fine, but great people are the key to winning. Every great CEO knows that, and great CEOs surround themselves with the best talent available.

Here are some of the questions we will answer in this book:

1. What are the qualities that CEOs want most from their CIOs, CTOs, and CISOs?

2. How do senior tech executives create an enduring culture of leadership that will sustain the enterprise over several generations?

3. What are the attributes needed to innovate in core, parallel, and new markets?

4. What's required for nurturing a true culture of innovation?

5. How do you accelerate transformation in the modern digital enterprise?

6. What are best practices for institutionalizing a security-focused mindset throughout the enterprise?

7. What are the qualities and attributes necessary for building world-class partnerships?

8. What are the best strategies for simplifying IT and removing complexity from systems?

9. How do you create a durable culture for achieving optimal speed to performance, speed to innovation, and speed to cash?

How This Book Is Structured

This book is divided roughly into two main sections. The first part (Chapters 1 through 4) features our signature interview-based approach to research. It is pure peer-to-peer knowledge transfer, based on one-to-one conversations with top executives and leaders representing industries and businesses all over the world.

The second section (Chapters 5 through 7) features deeper dives into the various challenges and opportunities facing modern CIOs and other technology executives. You will find both sections highly useful, readable, and filled with actionable ideas and insight gathered from the "best of the best" leaders in our industry.

Chapter 8 is "Key Takeaways," which summarizes the main lessons we learned from our research. It is essentially a "cheat sheet" of important knowledge we acquired during the process of writing this book.

Following the pattern established in my previous books, we have included short biographies of our expert sources at the end of *The CEO of Technology*. You'll find the bios themselves to be highly useful sources of information.

Chapter 1

Qualities of Courageous Leadership

Do You Have the Courage to Lead Your Team to Greatness?

Over the past 12 months, my research has focused on exploring the value of courage in leadership. At our CIO Executive Leadership Summits, we've spent hours talking about the qualities required for leading IT teams to greatness, higher levels of achievement, and superior performance.

We've come a long way since the days when we simply assumed that some people were born for leadership roles and others weren't. Yes, it still takes talent and the right temperament to become a great leader. But it also takes experience and the willingness to learn. You need an open mind and trusting heart to guide your IT team across today's landscape of unprecedented disruption and continuous innovation.

Our research points to five foundational pillars of courageous leadership:

1. Innovation
2. Security
3. Talent
4. Simplification
5. Trust

Innovation is the combination of invention and economic value. It's never just about new technology; there's always a business driver. Great CIOs and IT leaders understand that successful companies are fueled by continuous innovation. In highly volatile markets that are easily disrupted, innovation is the only proven strategy for beating the competition.

Security is absolutely critical; just look at the headlines if you doubt the importance of security in the modern enterprise. In the past, security was considered a cost of doing business. Courageous IT leaders are more likely to see security as a competitive advantage. If your security is better, deeper, and more agile than your competitor's security, that's a point in your favor. No companies are entirely safe from hackers and cyber criminals, but great IT leaders know how to balance security and agility to outmaneuver the competition.

Talent is the most important resource in today's ultracompetitive economy. You can never have enough talented people on your staff. Great leaders attract and surround themselves with the best people. It takes more than money to hire and retain top talent—you need to understand how talent works and what motivates people to excel in the twenty-first-century workplace.

Simplification has become the new normal for IT users at all levels of the enterprise. Even if when the underlying systems are incredibly sophisticated and complex, the user experience—both internally and externally—must seem natural, intuitive, easy, and painless. The modern workforce has no patience for complexity; it wants immediate results, in the plainest terms. That's why IT simplification is so critical; people will abandon systems and interfaces that require lots of effort.

Trust is the bedrock of courageous leadership. You cannot build a team without trust. Great teams depend on deep relationships that only grow when people trust each other. Great leaders understand the basic secret of trust: It develops when you trust other people, when you get out of their way and you empower them to do their finest work.

Michael Dell: 5 Tips for CIOs to Lead Digital Transformation Strategies

As chief information officers (CIOs) and other executives become more deeply involved with digital transformation, one thing that's clear to them is that digital transformation is not an IT project but is instead at the center of business strategy.

Without question, digital is driving the future of business. Consider this: By the end of 2017, revenue growth from information-based products is *projected* to be double that of the rest of the product/service portfolio for one-third of all Fortune 500 companies, according to IDC.

I recently spoke with **Michael Dell**, the legendary founder and CEO of Dell Technologies, one of the world's leading providers of technology solutions and service for accelerating digital transformation, regarding how digital transformation is unfolding along with recommendations for CIOs on how best to lead the digital charge.

Here's is a transcript of the insight and advice that Michael shared with me:

Hunter Muller: As the CEO of a major technology enterprise, what have you learned about the challenges of developing new product/service offerings and expanding into new markets?

Michael Dell: Listening to your customers is the key to surviving and thriving in business. At Dell, we call it having "big ears." From the beginning, we built the company from the customer back, partnering with them to design new products and services. In 33 years of business, customer feedback has proven to be the best crystal ball there is.

Equally important is a willingness to disrupt yourself. We're living in an unprecedented time. Data, the Internet of Things (IoT), and technology have replaced industrialization as the driving force of progress and transformation. To win today, you must be fearless about transforming for the future. That's something I've always embraced—whether it meant diverging from the business model that fueled our early success or taking the company private after 25 years of being public, or combining with EMC to become the largest enterprise infrastructure company in the world. Business as usual is how you *lose* in the digital era in which we live.

HM: From your perspective, what should every C-suite executive understand about digital transformation?

MD: I talk to a lot of business leaders from every industry and in every region. None are asking *if* they should digitally transform. They know their future depends on it. As a matter of fact, 45 percent of them believe their businesses may be obsolete in five years and nearly half don't know what their industry will look like in three. It's creating a lot of digital fear out there. But it shouldn't. When the CEO and board see digital transformation as a business opportunity rather than an IT problem, you are halfway there. Then it's about finding the right IT partner with the expertise and depth and breadth of solutions and services to get you where you want to be.

HM: Thanks to their technical expertise and unique view across the enterprise, CIOs are often deeply involved in digital transformation efforts. Can you offer recommendations as to how CIOs should be thinking and acting like CEOs to help lead digital strategies?

MD: The CIO role has changed because technology's role in business has changed. No longer relegated to the back office, technology is how modern enterprises are gaining competitive advantage, and this aligns the CIO and CEO like never before. A few pieces of advice for CIOs:

1. Digital transformation, like any key agenda for the CIO, must be driven from and closely linked to the core strategy of the company.

2. Embrace the fact that every business will to some degree be a software business in the future. So start building (or flexing) your software development muscle now to get ahead and stay ahead of the competition.

3. Focus on data. The truth is out there, but you have to get good at aggregating, integrating, analyzing, and acting on customer data. The CIO holds the key to this most valuable resource.

4. Create a culture of fearlessness when it comes to trying new ideas and technologies. It's never been simpler to pilot a technology solution that can dramatically change an organization's effectiveness or competitiveness.

HM: Despite all that's been written about digital transformation, many executives still don't really understand it. Which aspects of digital transformation are hardest to grasp, and what's the best way of getting executives to understand the strategic value of digital transformation?

MD: *Digital transformation* is a buzzword that leaves room for interpretation. But at its core, it's really about figuring out how to make the most of all this rich data that today's technology is cranking out at lightning speed. How do you use artificial intelligence, machine learning, deep learning, etc., to turn real time and older time series data into useful insights to make your products and services better and smarter?

When done right, it creates a continuous cycle. You embed sensors in your product (think Tesla or GE) or create a software to deliver a digital service (i.e., Uber or Spotify). The embedded software feeds you real-time data about product usage, performance, customer information, etc. You use those insights to quickly improve your product or service or to innovate new ones again and again. It's a virtuous cycle, and the faster you go through it, the better you get and the greater the distance between you and your competition. That's the definition of strategic value if you ask me.

HM: What additional leadership advice would you offer to C-suite executives that are striving to transform their businesses? What should they be thinking about?

MD: Don't be paralyzed by digital fear. Transformation is within reach of every business, large and small. There's this perception that the advantage is solely with the disruptors. But we're helping giant global retailers, automotive companies, etc., masterfully reshape and digitize for the future—and their scale is helping them do it.

Transforming IT for the Digital Age

Some companies have large IT budgets. Then there is JPMorgan Chase & Co.

The 2017 IT budget for the financial services giant is $9.6 billion. To help put this into perspective, the bank's IT budget by itself would place it at number 302 on the Fortune 500 list.

At the helm of the bank's IT organization is Dana Deasy, managing director and global CIO at JPMorgan Chase, which was recently named among Fortune's Most Admired Companies. Dana, who was recently featured in a keynote interview at HMG Strategy's 2017 New York Summit of America, shared his recommendations for successfully driving digital transformation.

One of the essential aspects of digital transformation for IT organizations is marketing the vision and execution effectively to stakeholders. Dana, who has overseen global technology at JPMorgan Chase for the past three years, said this is true even though marketing is rarely a strong suit for most IT groups. "Marketing needs to be a key ingredient in driving transformation," he said. "You need to brand your transformation, give it a name and give it an identity. This can help people buy into your vision. It's powerful how people will rally behind it when you do."

As CIOs move forward with digital transformation efforts, he also recommends picking the top four or five key performance indicators (KPIs) that matter most. These measures of success can include external perceptions of the company's digital transformation efforts by customers and investors.

"What are they saying about your technology? Are they connecting with what you're trying to accomplish?" he asks.

It's important to have change agents at all levels of the organization who know the pulse of your employees and how they're reacting to the transformation.

As a starting point to achieve this, he says, "you need to address the cultural barriers that may be holding you back."

Perhaps the greatest piece of advice that Dana Deasy offers to CIOs is this: Believe in your instincts.

"Whenever you try to drive change, everyone is going to give you advice," he says. "Take your own counsel, live with your decisions. Transformations are meant to be difficult and people will inject different opinions. Remember you were hired to be a change agent, and this includes your ability to make good decisions."

The Building Blocks of Digital Transformation

As companies move forward with digital transformation initiatives, organizational leaders are exploring a range of opportunities to obtain value. These include identifying new ways to reimagine the customer experience, opportunities for streamlining company processes, along with methods for revolutionizing work tasks and boosting worker productivity.

While many practitioners launch into digital by attacking the organization's biggest pain points, ultimately, it's important for decision-makers to use a wide lens to plot the digital strategy across the enterprise for maximum benefit. Doing so ensures that digital investments will be optimized and that digital initiatives won't be siloed between disparate functions or business units.

Taking an integrated approach to digital is key. According to IDC, by 2018, 70 percent of siloed digital transformation initiatives will fail due to insufficient collaboration, integration, sourcing, or project management.

Regardless of whether a CIO has ownership for digital strategy or is helping to guide these efforts, it's important for the CIO and IT team to orchestrate its role in digital execution with key stakeholders across the business, says Patty Hatter, SVP, Professional Services, McAfee.

"The most successful transformations are those where everyone is on the same page—both IT and business leaders—and applying their best resources towards a common set of goals," says Patty, an HMG Strategy 2016 Transformational CIO Award winner.

Whether the CMO, chief digital officer, or another executive is leading the digital charge, the CIO plays a critical role in aligning technology with business strategy. "It's an opportunity for the CIO to bring more to the table, to marry the IT, business, and functional skill sets and to identify the opportunities from a top- and bottom-line perspective," she says.

Another critical component for successful digital execution is identifying how processes need to be designed properly. In too many cases, decision-makers attempt to bolt digital technologies on top of legacy processes that aren't designed for digital workflows.

"New technology is great, but new technology on top of an old process doesn't buy you much of anything," Patty says. "That's where the CIOs and their teams really need to connect the dots. It's about how the bigger pieces fit together, how your company is working with its customers and its partners and how you streamline that whole ecosystem and lifecycle."

Because digital transformation also requires new skills across the enterprise, the CIO and IT team should talk to end customers as well as the company's business partners to help determine the skills that are needed to address their needs.

It's also important for the CIO and the IT team to stay connected with new technology developments to cultivate a firm point of view as to where markets are heading. Patty says, "You want both the business and the technology team to know enough about each other's perspectives to get on the same page and to identify the bigger opportunities that are out there."

She also isn't a big fan of taking a "big bang" approach with digital transformation initiatives. "With big bang, if you try to figure it all out in a conference room instead of testing it out with customers and employees, that's probably not a recipe for a high probability of success."

Instead, she recommends taking an agile approach to testing new ideas with different lines of business.

"Digital transformation is really a journey, a learned skill," Patty says. "The better a company gets at it, the faster you can move and the more risks you're able to take and succeed at."

Are You a Great Leader, or Just along for the Ride?

I had an excellent conversation recently with Zack Hicks, chief executive officer and president of Toyota Connected and chief information officer of Toyota Motor North America (TMNA).

In his role at Toyota Connected, he's reinventing the way we think about mobility, developing solutions that excite our customers and anticipate their needs, by utilizing advanced technology and through the art of predictive intelligence.

As CIO at TMNA, he drives strategy, development, and operations of all systems and technology for Toyota's North American operations. Zack is totally focused on aligning the efforts of business operations, strategic planning, and technology to drive business innovation and efficiency.

In our conversation, he emphasized the importance of taking an executive leadership role—and resisting the urge to just go along for the ride.

"Some CIOs expect leather-bound business strategies will be dropped on their desks, and all they have to do is execute on those strategies. But that's not how I look at my job," says Zack. "You need to get out of your office and talk to the people in your company. You need to understand how they work and what they're doing. You need to stop just being an ordertaker and start being an executive."

Over the years, Zack has aligned the IT organization with the business, making sure the IT team genuinely understands the goals and objectives of the company's business leaders. "The more we know about the business, the better we can deliver effective solutions," he says.

From his perspective, now is the best time to be a CIO. "Sure, you can get bogged down in all the legacy stuff... but you also have the opportunity to transform the business," he says. "If you're waiting for the business to hand you a flow chart, you'll be missing a great opportunity."

I think that's great advice. Take the bull by the horns and create your own destiny. Don't wait for someone else to tell you what to do—figure it out for yourself, and execute on a strategy you've helped develop.

Great CIOs are also great leaders. They don't sit around waiting for orders—they move full speed ahead, accepting the risks and reaping the rewards. Which kind of leader are you?

Empowering a Fearless and Purpose-Driven Team

Naresh is no stranger to acquisitions, joint ventures, and spinoffs. A few of HP's more notable transactions include its 1999 spinoff of Agilent Technologies, the sale of its Medical Products Group, the 2010 purchase of mobile device maker Palm Inc., and its split into two companies—HP Inc. and Hewlett-Packard Enterprise. Naresh Shanker was at the center of each of those transactions.

The formation of HP Inc. has presented yet another opportunity for Naresh to spread his wings and demonstrate powerful IT leadership. He was previously CIO at Palm Inc. and CIO of HP's Printing and Personal Systems Group.

"Working with a core team of individuals over two decades across multiple transactions and industries has helped me execute strategies that meet the demands of a rapidly evolving market," he said, when talking about the importance of developing an effective team equipped with the right skills, drive, and stamina to successfully support transitions of scale.

Naresh says that when talking about leading large teams through mergers, acquisitions, or divestitures, a lot of myths around leadership and change come apart when transactions are complex, involving more than 100+ countries, 1,000+ applications, and 20,000+ servers. It is not about the size of the team but your ability to integrate the individuals within and across global teams.

"As an example, I have worked with talented solution architects who are leaders because they have been able to bring holistic thinking into solving complex problems across highly matrixed organizations. These qualities become even more critical to the success of an endeavor, especially when working the seams in dynamic and complex environments that traverse multiple geographies," he says.

"I look at the true value employees can create regardless of the hierarchy of their job or role," says Naresh. "It is important to foster an environment that enables people to excel—whether by leading a team or by working as a unique, individual contributor. Coaching, mentoring, and providing skills training have been my focus as I have taken teams through demanding transitions," ensuring that the individuals and the team or the company not lose focus on the contribution of the individuals.

One of the key attributes of great leadership is the ability to lead the narrative that effectively connects people, especially when managing an IT team that supports diverse businesses and go-to-market notions worldwide.

"It's about being able to communicate and internalize a shared vision and purpose. Because of the 24x7 nature of our work, it's important for every team member to be able to proactively take the baton forward based on their clear understanding of shared goals. You've got to create a highly collaborative and trusted environment to inspire commitment, especially when stakes are high and potential burnout is severe."

His core team has been with him through several major corporate merger, acquisition, and divestiture transactions that have transformed diverse industries and customer experiences. Naresh says, "It's critical to create an ecosystem where people are allowed to play fearlessly and with a purpose."

Naresh also recognizes the importance of fostering a culture that rewards success while acknowledging failure and seeing a firm path to root-cause and corrective-action feedback loops.

SAP Relies on Culture of Innovation and Excellence to Stay Ahead

Peter Drucker famously said, "Culture eats strategy for breakfast." From my perspective, SAP, the world's leading maker of enterprise software systems, demonstrates the essential and inescapable truth of Drucker's immortal words.

I spoke recently with Helen Arnold, the CIO of SAP. Helen is also a member of the SAP Global Managing Board and serves as chief process officer (CPO) for SAP SE. Throughout our conversation, Helen emphasized the critical importance of establishing and nurturing a culture of innovation, inclusion, and operational excellence. I was deeply impressed by her passion and her commitment to serving the culture of a truly legendary multinational corporation.

"These are the most exciting times to be a CIO," says Helen. "I can tell you each individual day is really exciting. Our goal is leveraging technology to help our customers succeed in today's fast-changing and disruptive business environments."

I asked Helen to list the top challenge facing twenty-first-century CIOs. Her response was energetic and direct: "You have to be able to walk and chew gum at the same time. For IT, that means renovating the core systems while growing new capabilities and innovating with newer technologies. To accomplish that, you need a clear vision that supports the business strategy of your company."

In the past, the CIO focused mostly on running systems and finding the best tools for the business. Today, Helen says, the CIO is an indispensable innovator and trusted partner to business units across the enterprise. "You have to be extremely committed and dedicated to the business strategy," says Helen. "You need the ability to defeat the complexity around you and deliver real value to the people who depend on you."

For the modern corporate CIO, strong and effective leadership is absolutely fundamental. "Because there is so much disruption and change today, being a leader is more important than ever," she says. "Business strategy relies on a strong culture of shared values and common goals. Everyone at every level of the enterprise needs to understand the strategy, not just a handful of people. That's why we focus on our culture. If we make a promise, we keep it. We deliver. That's the key. Being CIO is more than just a title on a business card."

I was genuinely inspired by my conversation with Helen. She embodies the combination of business knowledge, technical expertise and strong leadership skills required to succeed in today's hypercompetitive markets. I'm glad that Helen took the time to share her insights with our team at HMG Strategy, and I look forward to future conversations with her.

The Modern CIO Is a Tech Investor, Not Just a Tech Consumer

I had an excellent conversation with Clark Golestani, president, Emerging Businesses, and global CIO at Merck, a leading global biopharmaceutical company. Merck has been bringing forward medicines and vaccines for many of the world's most challenging diseases through its prescription medicines, vaccines, biologic therapies, and animal health products.

Clark and his team embody the Merck tradition of invention, safety, and excellence across multiple disciplines and dozens of markets. Additionally, Clark represents a new kind of CIO—an experienced executive who identifies and invests in new technology that spurs innovation across the enterprise.

"The CIO community has done a great job of driving operational excellence," says Clark. "Now we've got to take the next steps, which involve being more engaged with the startup community. We've spent a lot of time learning how to be great partners and strategists. Now the challenge is learning how to become great investors."

I genuinely appreciate how Clark defines the challenge and the opportunity for CIOs and senior IT leaders. I agree completely that it's time for us to learn new skill sets and to become more deeply engaged with the startup world. It doesn't necessarily mean that every CIO will become a VC; but every CIO needs to begin thinking like a VC. Moreover, CIOs also need to start thinking—and acting— like CEOs.

"CIOs need to think like CEOs and actually run IT like a business," says Clark. That means focusing on earnings and revenue, not merely on costs. "Great CEOs don't go to their boards and say, 'I need more money to succeed.' Great CEOs work with what they have to achieve success."

I firmly believe that Clark has articulated the main difference between the mindsets of traditional CIOs and modern CIOs. Traditional CIOs are mainly consumers of tech, while modern CIOs assume active roles in identifying, nurturing, and developing new tech. It's a huge difference in worldview.

We're fortunate to have a wonderful global community of innovative and inventive IT vendors. But as CIOs and IT leaders, we're also responsible for carving out our own destinies and making sure that our companies succeed. From my perspective, that means taking a more active role in the development of new technologies and new techniques for achieving our strategic business goals. It's our responsibility, plain and simple.

Taking Your Seat at the Table with the Board

The role of the chief information security officer (CISO) has evolved in many ways over the past decade. For instance, as the threat landscape has continued to change and become more complex, CISOs have needed to improve how they communicate the nature of these threats and the response plan that's in place to address such risks to the C-suite and the board of directors.

Plus, as cyberthreats have become more widespread and as the cybersecurity talent shortage has become more acute, CISOs have had to become much more creative in their approaches to identify, recruit, develop, and retain cyber professionals.

But perhaps the most significant change to the CISO role is that it has become considerably more business-focused in recent years. While CISOs still need to be technically competent, they also must be able to communicate the company's security posture, its response to information security threats, along with its risks, mitigation, and controls in business terms that the C-suite and board of directors can understand.

Not only are members of the C-suite and the board looking for cybersecurity to be couched in business terms, but they sometimes need to be steered away from their view of cyber concerns as a technical issue.

According to a 2016 survey of Deloitte's CISO Labs participants, 79 percent of information security leaders indicated that they were "spending time with business leaders who think cyber risk is a technical problem or a compliance exercise."

"I'm being asked to be more of a business leader, to respond to executive management and board interests in the topic, and to communicate differently with the board than we had to a few years ago," says Michael Wilson, SVP, CISO McKesson IT, McKesson Corporation.

One of the ways that Michael is communicating the company's security posture is by arming members of the board with dashboard tools that enable them to track how effectively McKesson is tracking with cyber security governance, protection, response, and recovery efforts.

"I use these tools as well. It tells a story about our organization's maturity," he says.

Meanwhile, cyberreadiness updates with the board and with McKesson's audit committee have also become more frequent in recent years. "The cadence is up, and the concern is there. Most boards are struggling to have IT representation and now we have the security piece which brings it to another level."

Not only are his discussions with the board and C-suite business-focused, they're also concise, Michael says. "The communication with the board is short—you don't have a lot of time. They're looking for metrics to be consistent as they've seen in other places. What is the threat posture, what are we doing well, what are the gaps, and what is the plan to address those gaps? These are the things they want to know."

Even as McKesson is moving forward with digitizing its various businesses and embedding security into its digitized operations, Michael Wilson is finding that his role and communications has become increasingly business-focused and less technically-oriented.

"I'm being asked to be more of a business leader and to respond to interest on cybersecurity topics and to communicate differently than I had to a few years ago," he says. "It's become less of a technical role and more about balancing our cyber security needs with our business strategy."

When Pitching New Ideas, Bring Prototypes and Working Demos

I had the opportunity to sit down with a legendary technology leader in Silicon Valley recently, and I sought his advice on the best ways for preparing for meetings with the C-suite and board of directors.

Specifically, I asked him to describe how CIOs and other technology leaders can make the most of their meetings with senior executives.

His answer was a wonderful blend of insight and practicality. Basically, he advises CIO to *show*, rather than *tell*.

A picture might be worth a thousand words, but a working prototype is worth even more!

In today's world of agile development and rapid prototyping, it's easier than ever to create a working version of the idea or concept you're trying to explain in a meeting.

"You can build that first embodiment of the product or the app," he said. "Before even walking in the door, you can build the first version. Then instead of talking, you're showing."

Nowadays, more senior executives and board members are familiar with the principles of agile development. That means you can bring an MVP (minimum viable product) version of your idea to the meeting.

"Then you can walk into the meeting and say, 'Look, here's what we did using what we have today and here's what we can do if we combine our present-day capabilities with what we'll have in the future.' That's a very powerful statement," he said.

He also recommends that CIOs network regularly to keep up with the competition. "If you're the CIO at State Farm, you want to know what they're doing over at GEICO. If you're at FedEx, you want to know what they're doing at Amazon. Then you're walking in with a prototype and with competitive intelligence. That's a powerful combination."

My friend is truly one of the most brilliant tech leaders I've met. His advice is incredibly relevant to all of us in the tech industry, and I urge everyone to keep his words in mind as you prepare for your next meeting with the board.

Understanding the Difference Between Innovation and Disruption

I had a highly useful conversation recently with John Rossman, a managing director at Alvarez & Marsal. A former Amazon executive, John is a successful author and speaker. His books, *The Amazon Way: 14 Leadership Principles Behind the World's Most Disruptive Company,* and *The Amazon Way on IoT: 10 Principles for Every Leader from the World's Leading Internet of Things Strategies*, are considered must-reads for twenty-first-century entrepreneurs, executives, and investors.

We spoke about the need for technology leaders to understand the difference between *innovation* and *disruption* in the context of the modern global enterprise.

"A lot of people use the word *innovation* because it's a happy word, as opposed to *disruption,* which is a messy word," John says. Disruption is tough. It's risky and unpredictable. You can't guarantee the outcomes. Jobs change and people feel uncomfortable."

For many executives—including CIOs, CTOs, CMOs, CFOs, CEOs, and board members—maintaining the status quo is preferable to risking an uncertain future. That simple fact is a big part of the challenge.

As Clayton M. Christensen notes in *The Innovator's Dilemma*, disruption isn't a straightforward or one-dimensional process. It's complicated and multidimensional. That's why many executives find it unsettling.

John says, "A classic example of disruption is transitioning from a product organization to a services organization where you're selling outcomes instead of tangible goods. That requires a transformation of your business model, and the impact will be felt across every function in the organization."

Many organizations find it easier to follow a path of incremental innovation. "There can be a lot of value in incremental innovation, whether it's a process improvement or a digital transformation," he says. "But even when you're innovating incrementally, there are huge challenges to overcome."

John Rossman's insights and advice mesh nicely with my theme of courageous leadership, since transformation and disruption aren't strategies for the meek or timid. "Patience is an important quality, but you've got to remember that disruption is a high-risk game. These are the moments when it's not about management, it's about leadership. Leadership matters when you are making fast, dramatic, and scary changes."

I really appreciate how he sets the stage for a deeper and more comprehensive conversation about the interplay between innovation, disruption, risk, transformation, courage, and leadership.

Rethinking the Concept of Bimodal IT

I had a great exchange with Pat Gelsinger recently and I wanted to share some of his insights. Pat is the CEO at VMware, a global leader in cloud infrastructure and business mobility, and his opinions are widely respected across the technology industry.

Our exchange touched briefly on the topic of bimodal IT, and Pat's thoughts on the matter inspired me to revisit my own thinking on the subject.

As most of you know, bimodal IT is a concept that essentially separates IT into two functions. One function is responsible for operating and maintaining the company's existing systems, while the other is tasked with developing new and innovative solutions.

Ironically, the problem with the bimodal concept is two-sided: One, it divides the IT team. Two, it overlooks the amazing capabilities of virtualized infrastructure.

I wholeheartedly agree with Pat's idea that virtualized infrastructure is inherently agile infrastructure. From my perspective, having virtualized infrastructure means you're ready for innovation.

In other words, when you have virtualized infrastructure, you don't really need bimodal IT. Furthermore, I think it's fair to say that bimodal IT is an idea for making sure that at least some IT resources are dedicated officially to innovation. But it is just an idea, not a framework or a full-fledged strategy.

It is also quite possible that bimodal IT would become a distraction, since it would add a new layer of administration to the IT function and potentially create unnecessary competition between members of the IT staff.

"While it isn't inherently wrong to argue for resource sharing across old and new systems, in practice, a bimodal IT strategy of siloing and sequestering IT teams pits 'old and slow' legacy teams against the 'new and cool' team of innovators. This perceived dichotomy reinforces discord, budget conflict and complaints of organizational favoritism," Pat wrote in a recent op-ed post for the Morning Consult.

There's no question in my mind that digital transformation often requires dramatic organizational changes. It also often requires new skills and fresh talent to develop and execute on a digital vision and adapt to a digital mindset. But the key to digital transformation isn't purely organizational. The technology piece is also critical, and that's precisely why having a virtualized infrastructure is essential to digital transformation.

Learning Valuable Lessons about Collaboration from the World's Largest Staffing Firm

Acquiring new skills and capabilities always requires moving out of your comfort zone. If you're totally comfortable in a new situation, then you're probably not learning much.

I learned several valuable lessons from my recent conversations with Bernadette Rotolo, former group vice president, Global Head of Solutions Delivery Management, focused on digital offerings at Adecco Group, the world's largest staffing and HR services firm.

Here's some interesting background information: With more than 32,000 FTE employees and around 5,100 branches in over 60 countries and territories around the world, Adecco Group offers a wide variety of services, connecting around 700,000 associates with clients every day. The services offered fall into the broad categories of temporary staffing, permanent placement, career transition and talent development, as well as outsourcing and consulting. Based in Zurich, Switzerland, the Adecco Group is a Fortune Global 500 company.

When an enterprise is large and successful, it can be hard to nudge it out of its comfort zone. But like other truly great companies, Adecco isn't afraid to change with the times. "As markets evolve, business models need to change," says Bernadette.

Rather than fighting it, Adecco learned to adapt and evolve to keep up with changes in the marketplace. While many staffing companies view social media platforms as competitors, Adecco decided to partner with social media companies and learn from them. Adecco wisely followed the old saying, "If you can't beat 'em, join 'em."

Bernadette understands the power of harnessing partnerships in the modern global economy. "Matching people and companies is both an art and a science. And it turns out the process is similar to dating in many ways." Just as Peter Drucker famously spoke about culture eating strategy for breakfast, poor matches between companies and their leaders will stunt growth and innovation. It's important to get the right people in the right places for the right purposes.

I looked forward to learning more from future conversations with amazing thought leaders like Bernadette and other visionary technology executives who understand that being successful means stepping out of your comfort zone.

Great CIOs Strive to Emulate Great CEOs by Focusing on Essential Leadership Skills

One of the top challenges facing most CIOs is leadership, plain and simple. Take a close look at any great CEO and you will see a great leader. That's the honest truth, and it's not likely to change. Great C-level executives are almost always great leaders first.

My friend John Foley has spent many years in leadership roles. He was a former lead solo pilot of the Blue Angels, the U.S. Navy's high performance flight demonstration team. As a career Navy officer, jet fighter pilot, and member of the legendary Blue Angels, John has studied the art and science of leadership from a unique perspective.

I asked him to describe the traits and characteristics of a great leader. Here's a snippet from our conversation:

"Great leaders inspire. They recognize potential in the people around them. Great leaders are always creating opportunities, for themselves and for the people on their teams," John said. "Great leaders are also calm under pressure. A wonderful example of a great leader is Greg Wooldridge, my "boss" on the Blue Angels. Even in times of extreme stress, Greg's calm and confident voice on the radio kept us focused on the tasks at hand."

I'm grateful to John for sharing his insight, and I really appreciate how he describes those essential characteristics of excellent leadership. Notice how all three of the leadership traits are inherently altruistic and benevolent. Great leaders focus on helping their teams achieve success and continually improve their performance. Great leaders stay calm and reassuring, even in the most difficult circumstances.

I believe that CIOs can learn valuable lessons from world-class leaders in all walks of life. As I've written and said many times, successful CIOs need more than superior technology skills—they need to acquire the characteristics and behaviors of great leaders.

Chapter 2

What Keeps the CEO up at Night

Modern CEOs Want Dependable Partners

Great CEOs typically wake up at night worrying about how their companies can stay ahead of the competition by innovating new products, opening new markets, expanding traditional markets, growing revenues from sales, increasing operational efficiencies, and improving the bottom line. It's a lot to worry about!

But in the final analysis, CEOs are judged by how well their companies perform in competitive markets. If your company isn't among the leaders, your tenure as CEO will likely be short, and not particularly sweet. That's why CEOs wake up at night and worry about how well their companies are doing compared to the competition.

Why should modern CIOs be concerned about what's keeping their CEOs up at night? The answer is simple: To a far greater degree than ever before, the CEO's strategy for success depends on the CIO's ability to deliver technology solutions that create real competitive advantages for the company. The days when the CIO's "customers" were mostly internal business users (aka "captive users") are over.

Today, the CIO's "customers" tend to be the same as the CEO's "customers." When the CEO talks about *customers,* he or she is talking about paying customers who generate actual revenue for the company. The fact that CEOs and CIOs now have the same customers represents a monumental shift of epic proportions, with major implications across the entire enterprise.

In addition to tactical issues, it presents a host of strategic challenges for the CIO. Enabling and executing the CEO's strategy for winning in competitive markets is fundamentally different from making sure that everyone in the company has an email address. For the CIO, the stakes are much, much higher than ever before.

That's the reason why the CIO's prime responsibility is making sure that he or she is totally aligned with the CEO. In the modern enterprise, there is no appreciate difference between business strategy and IT strategy. If those two strategies aren't precisely aligned, the enterprise will not achieve its goals and the board will begin looking for new senior executives to get the job done right.

In a recent post in *Information Age*, Ben Rossi summed it up brilliantly:

> *The role of the CIO will become will be an increasingly important one. With their knowledge of a company's existing IT infrastructure and future demands, it will be down to CIOs to take the lead and manage a company's digital transformation strategy.*
>
> *Digital transformation is touching all aspects of business, and as a result all enterprises need to be aware that every system upgrade, connection or native and third-party applications added to existing IT infrastructure, on-premise or in the cloud, will increase service delivery complexity, scale and operational risk.*

From my perspective, that certainly suggests a new and deeper partnership between the CEO and the CIO.

For the first time in history, the fates of the CIO and the CEO are fully intertwined, intermeshed, and interdependent. Whatever keeps the CEO awake at night should keep the CIO awake, too.

What Does It Mean to Be the Chief Digital Officer of a Major Global Enterprise?

Bill Ruh is the chief digital officer of GE and CEO of GE Digital, the organization that unifies GE's digital capabilities and integrates them into one strategic group. GE Digital brings together a variety of resources, including software development teams, global IT (information technology), and OT (operational technology) security solutions.

We recently asked Bill to talk about his new role and explain how he sees his duties fitting into GE's overall strategy. Here are excerpts from an email he sent in reply to our questions:

"As CDO, I'm focused on accelerating the transformation of GE into the world's premier digital industrial company," writes Bill. "Part of my job is working closely with the executive leadership team as we collaboratively write the playbook for the new digital industrial economy."

Another element of Bill's job is making absolutely certain that all the various parts of the GE universe fit together to create a seamless operating framework for delivering value to its customers.

"It's a challenge, no question," he writes. "From my perspective, the job is about integration and alignment. My main task is focusing our deep resources and talent into one laser-sharp beam that will change the way the world does business, forever. It sounds like a tall order, and it is."

A quick look at the stock market shows the rapid ascent of digital companies, and the steady decline of many traditional industries. There's no doubt that GE is betting heavily on the growth of a global digital economy.

"We're at one of those great moments in history when a brighter future seems within our grasp," writes Bill. "If we do this right, the digital industrial economy will bring progress and prosperity to billions of people. That would be genuinely transformational, and I'm confident that we can do it. We have the talent, the experience, and the knowledge. The creation of GE Digital gives us the formal structure we need to continue learning, building and growing."

I look forward to watching GE's progress as it builds the groundwork for a new digital ecosystem that will genuinely transform our lives. It seems clear that we're heading into a new age that is increasingly digital and hyper-connected. Thanks to Moore's law, our devices are becoming smaller, more powerful, and less expensive. The big question is whether businesses can transform themselves quickly enough to leverage the advantages created by all of our incredible technology.

Visionary global companies such as GE, Intel, Microsoft, AT&T, Cisco, and HPE are pushing the boundaries and blazing trails into an exciting future in which hardware, software, and commerce all blend into a holistic digital economy. I'm confident that CIOs will play major roles in the ongoing digital transformation and that IT organizations everywhere will benefit from the realization that every business is a digital business.

Digital Transformations Will Echo Far and Wide

The rebranding of GE as a large-scale digital enterprise will resonate deeply across the IT industry. We're going to start hearing about new ways of blending hardware and software to create value. Predictive maintenance doesn't sound very exciting, but it's likely to have a lasting impact on our companies and their customers.

According to Bill, predictive maintenance will be a key driver of growth in what some people are already calling the Age of the Industrial Internet. Predictive maintenance is a combination of advanced analytics, materials science, low-cost sensors, and global information networks. In essence, predictive maintenance translates into zero unplanned downtime. Imagine a world in which no device or system ever fails unexpectedly. That's the promise of predictive maintenance.

In a recent blog post, Bill explains why the idea of zero unplanned downtime will fundamentally change business: "Knowing in advance when a vital component is likely to break means you can replace it on *your* schedule—which is much safer and much less expensive than being forced to replace or repair something in the middle of the night or during a hurricane."

The blending of IT and OT creates incredible new efficiencies, with the potential for transforming virtually every industry, in every corner of the world. For CIOs, that means more transformation is ahead. It also means more opportunities for IT to demonstrate its value to the enterprise.

"When we significantly reduce operational costs, we open the door for a whole new generation of products and services that would have been far too expensive in the past,"

Bill writes. He also raises the fascinating idea of applying "long tail" economics to industrial scenarios. The "long tail" concept means that not every product has to be an instant "bestseller" to generate long-term profit. "I find this idea extremely exciting, because it elevates predictive maintenance into a force for economic transformation," writes Bill.

GE: An Industry Giant Continues to Transform and Evolve

Like many of us, I'm deeply interested in GE's continuing transformation, and I'm confident the company's management team is heading in a good direction that will launch a wave of new business opportunities across the IT industry.

After my first round of interviews with Bill, our research team followed up with a list of questions about GE's remarkable metamorphosis from an industrial giant to a major force in digital technology.

Here are the questions, followed by Bill's responses:

1. **How has GE Digital helped drive the transformation of GE from a traditional to a digital enterprise?**

    We really started this digital transformation journey simply to find a way to help make GE as an industrial company more productive. We call it GE for GE. On that journey, we discovered that there really wasn't the right solution built for the unique needs of industrial companies. We developed a solution to make our services business more efficient and have seen tremendous results. Last year alone we saw $730 million in internal productivity savings. To ensure we were truly bringing the entire company along on this journey, we created a new leadership model with chief digital officers (CDOs) for each of our businesses, establishing an L-shaped structure that allows us to leverage our scale, coupled with digital DNA.

2. **How were lessons learned in one GE unit applied to create value in other GE units?**

FieldVision is a great example of a solution that we have deployed internally to drive productivity and have been able to create value by applying it to other parts of our business. FieldVision was deployed as a way to modernize our field operations, providing a mobile field services application to our teams leveraging Predix, in order to simplify the way people work, taking out non–value-added activities that they are doing in their job day to day and automating as much of the work as possible, but then providing one view to how they do their job. We've seen $200 million of productivity gained through delivering a truly digitized field technician workforce, which we've been able to scale across our Energy Connections, Power, and Oil & Gas businesses.

3. **How have the role and responsibilities of the chief digital officer evolved, and where is it heading?**

I am seeing more and more companies not only create a CDO role, but having IT report into that role to drive real business value through technology and digital solutions. This is something we've done within GE with the CEO of our Transportation business being a former GE CIO.

4. **How can the digital twin concept be applied to many industries and business segments?**

The digital twin is the bridge between the physical and the digital world. Machine learning and AI coupled with modeling and data allow you to understand past and present operations, and make predictions about the future, driving asset productivity across the product lifecycle. We have hundreds of thousands of digital twins active today—ranging from aviation to improve fidelity, detection rates, and repair accuracy to power plant operations where we can improve reliability and reduce production costs. Digital twin technology has the power to span across all industries where the value is in assets. We foresee this same technology applying to the human body, changing the face of healthcare as we know it.

5. **How has software development and hardware design evolved to meet the demands of customers in rapidly changing markets?**

This is really at the core of what we're doing at GE Digital; we are the leading software company for the Industrial Internet. We've taken our industrial hardware and merged it with software, apps, and analytics to enable faster, smarter, and more efficient operations. We're pioneering technologies to help companies capitalize on the Industrial Internet, fueling productivity and value from existing assets and enabling new business models and growth potential. The digital twin is key to fully digitizing the physical world, creating a digital model for every physical asset to understand how equipment will perform and present options to extend an asset's life for better business outcomes.

The Shift to Digital Business Creates Incredible New Opportunities for Collaborative CIOs

I had a thoughtful conversation recently with Lee Congdon, the CIO at Ellucian, an education technology company serving 2,400 institutions and 18 million students in 40 countries. The former CIO at Red Hat Software, Lee has more than 25 years of experience as a technology leader, and his perspective is shaped by decades of hard work on the front lines of IT transformation.

Before joining Red Hat, he was managing vice president (VP), information technology, at Capital One, where he developed and delivered IT solutions for the firm's corporate functions and Global Financial Services group. Prior to Capital One, he was senior vice president (SVP), strategic initiatives, at Nasdaq, where he led the organization's efforts to identify, implement, and operate technology solutions for Nasdaq Japan, Nasdaq Europe, and other strategic global ventures.

"The world is shifting from an Industrial Age to an Information Age. It's no longer enough to produce great products. Today, companies need the skills and capabilities to connect with their customers. They need the ability to manage their supply chains in real time. Traditional companies are becoming digital companies, and that's a fundamental change," says Lee.

Companies are also struggling to embrace processes and technologies that help them make better business decisions, often in collaboration with their customers and stakeholders. For CIOs, that fundamental shift requires new skills. "The new skills aren't replacing your old skills, because you still need those old skills to run your IT organization," says Lee. "But you also need to develop new sets of visionary skills that weren't part of the job description."

Part of that new skill set involves collaborating with your company's business leaders in different ways than before. It often means transitioning out of the role of order taker and transitioning into the role of trusted technology consultant to the enterprise.

The cloud's ability to deliver new IT services easily and quickly creates both new opportunities and new challenges for the twenty-first-century CIO. "At minimum, the CIO has to deliver more value than what the business can do for itself," says Lee. Smart CIOs will position themselves as partners to the business, and will leverage their technical expertise to help the enterprise achieve meaningful business goals.

"It's a tremendous opportunity for IT organizations because they are one of the few parts of the typical enterprise that sees the entire business, sees all of the business processes. Many of the other parts of the organization simply don't touch the organization to that degree of depth," says Lee. "In some cases, the IT folks fundamentally know more about the business processes than the business decision-makers because they've implemented the systems and understand them."

I really appreciate how Lee Congdon frames the opportunity, and I especially like his vision of the CIO as a trusted technology consultant to the enterprise. Clearly, the modern enterprise will be increasingly reliant on a digital backbone, and that creates a world of opportunity for CIOs who can collaborate with their peers in the C-suite to generate real business value.

Understanding How Technology Drives Business Growth Is Critical to Market Success

Timothy Kasbe is chief information and digital officer at The Warehouse Group Limited, one of the largest retail groups operating in New Zealand.

Previously, he was COO at Gloria Jeans, a role that gave him a truly unique perspective on the absolutely critical value of technology in the twenty-first-century business enterprise.

"It's is a technology business that sells fashion," Tim says. As a former CIO, Tim keenly appreciates the many ways in which technology can be leveraged to drive growth and value. "Great companies understand the strategic value of information technology. American Express, for example, is a technology company that sells financial services. Tesla is a software company that sells cars."

Uber is widely admired for using technology to create a platform for exponential business growth. "Uber has never produced a single car. They don't own any taxis. And yet Uber is completely disrupting the transportation and taxi industries," Tim says.

Netflix has followed a similar trajectory. Long before it began producing great shows like *House of Cards* or *Orange Is the New Black,* Netflix destroyed the video rental industry by creating an innovative business model built on great technology.

Leveraging technology as a business generator and disrupting industries with new technology-supported platforms are two of three key trends driving innovation across the global economy, says Tim. "The third major trend is that fueled by the explosion of data. Combining data and algorithms allows you to create new products and offer new services very quickly, based on real consumer trends," Tim says.

With newer data technologies and advanced analytics, companies can respond to changes in the market at lightning speed and meet the needs of their customers before the competition figures out what's going on. "In the apparel industry, for example, the traditional product life cycle was about eight months," Tim explains. "We've shortened the cycle to six weeks."

Tim is definitely part of a new generation of executives who deeply understand the role of data and advanced analytics in driving strategic growth. His combination of skills and multidisciplinary background make him an exceptional player in a rapidly evolving global industry; his "comfort zone" includes business and technology.

From my perspective, Tim is an ideal role model for the modern corporate executive, straddling two worlds with supreme confidence and skill.

Driving Secular Change in the Modern Enterprise

Listening to Asheem Chandna of Greylock Partners describe the future is like riding in a time machine. As a pillar of Silicon Valley's investor community, Asheem has a privileged role in partnering with entrepreneurs to help create the future.

"The mobile phone has changed the game, and every business is going digital," says Asheem. "It's driving a huge secular change in how companies engage with customers."

For CIOs, that means shifting focus from "systems of record" to "systems of engagement." In the past, CIOs could afford to spend most of their time solving internal technology issues and managing internal "customers." But that's no longer the case, says Asheem. The company's customers are now the CIO's customers as well.

The rapidly expanding universe of digital devices and platforms creates thousands of new channels for interacting with customers. In the past, customer interactions were limited by time and distance. Now those constraints have largely vanished, thanks to the ubiquity of digital technology and broadband networks. Customers now expect companies to interact with them anytime and anywhere.

"You can be in touch with customers continuously, through all of these new channels of engagement," he says. "If you don't adapt to this new way of doing business, someone else will come along and capture your customers."

I agree totally with Asheem's assessment that we're seeing a secular trend. "In the old days, we spoke about the big fish eating the little fish. Today, it's the fast fish eating the slow fish," he says. The "new normal" puts tons of pressure on CIOs to deliver practical technology solutions. It also creates incredible opportunities for smart and agile CIOs to provide courageous leadership through turbulent times.

As an investor and entrepreneur, Asheem genuinely understands the deep value of visionary leadership. Greylock Partners has invested in new consumer and enterprise leaders such as Airbnb, AppDynamics, Cloudera, Docker, Dropbox, Facebook, LinkedIn, Palo Alto Networks, Pure Storage, and Workday. Clearly, Greylock is keeping a sharp eye on the future.

I'm delighted that Asheem agreed to share his keen insight and knowledge with us. Many of us try to predict the future, but Asheem seems to have a special gift for seeing around the corner and over the horizon.

Are Tech-Savvy CEOs a Help or a Hindrance to the Modern CIO?

Here's a phenomenon that has added a new dimension of complexity to the CIO's role: The modern CEO understands technology far better than his or her predecessors.

In fact, it's increasingly rare to find a CEO who does not possess a fairly good working knowledge of current digital technology. The typical CEO understands how technology creates value—if not precisely, then generally—and believes deeply that technology is absolutely essential to the success of the twenty-first-century enterprise.

It's all a far cry from the 1980s and 1990s, when many top executives regarded IT as a backroom operation.

This raises an interesting question: As CIOs, are we truly ready to deal with CEOs and other senior executives who understand technology?

I think we can all agree that it's better for the CIO when the CEO and the board understand the value of technology. But how should the CIO respond when tech-savvy executives second-guess the IT team's recommendations?

There's an old saying that goes, "A little knowledge is a dangerous thing." Sometimes that can apply to senior executives who know a little about technology.

So, what's the best way for CIOs to handle conversations about technology with their peers in the C-suite?

My suggestion would be to steer the conversation from technology to value creation.

Shifting the focus to value creation means you're talking about results. That will immediately make the conversation more productive and less argumentative. In reality, the conversation should always be focused on results, and not on the technology used to achieve the results.

I'm all in favor of CEOs who understand technology. But a big part of the CIO's role is reminding the CEO and other senior executives that technology is only a means to an end. The results are what matter.

Naturally, you'll have to figure out a diplomatic way of expressing all of that. But as the CIO, you are responsible for guiding the conversation away from technology and toward value creation.

Five Critical Qualities of World-Class CIO Leadership in the Modern Global Enterprise

In addition to the traditional role of technology leader, the modern CIO is expected to serve as a value creator and pioneer of genuine business growth in the modern enterprise.

For some CIOs, stepping up to a real corporate leadership role will not be easy. The good news is that leadership can be studied, practiced, and perfected—assuming, of course, that you're willing to make the effort.

The idea that great leaders are born to the role has been largely disproven. While it's true that some individuals might possess higher levels of charisma than others, all the "secrets" to successful leadership can be learned and sharpened over the course of your career. Leadership is definitely an acquired set of skills, and not a genetically inherited trait.

My extensive research into leadership has convinced me that successful leadership strategies are multidimensional, consistent, and built for the long term. Moreover, all the important leadership qualities are rooted firmly in traditional "people skills," such as empathy, awareness, collegiality, communications, and collaboration.

Over time, I have observed five distinct capabilities required for world-class IT leadership and continuous innovation in the modern globally connected enterprise:

1. *Multidirectionality.* The modern leadership model takes multiple paths and explores multiple options. It combines internal and external resources. It has focus and structure, but it is also flexible and resilient. It assumes certain levels of risk, with the understanding that risks are proportional to rewards, and therefore necessary for success in competitive markets.

2. *Inside / outside balance.* Successful leadership strategies leverage a blend of internal and external resources to find creative solutions and serve new markets.

3. *Redefined teams.* Successful leadership is all about building functional teams. In the past, team members

were selected for compatibility and basic skills. That's no longer sufficient—today's teams must include insiders and outsiders, people who can find and leverage the appropriate resources (whether external or internal), people who bring different views and opinions—people who might not even be considered "team players."

4. *Deep knowledge and market awareness.* Great leadership also requires deep and extensive knowledge and awareness of the competitive landscape. You have to know what the competition is doing and know your competitor's business—even better than the competition knows its own business!

5. *Partnering with the world.* The best leaders know that everyone is a potential partner. You must find good ideas wherever they are, and figure out how to make them work within your organization to create new value for customers.

This is by no means a complete or exhaustive list. We'll add more key qualities and capabilities as we continue researching and exploring the best ways for IT leaders to create value and bring innovation to the modern enterprise. We're in the opening innings of a long game. The outcome will depend largely on the level and quality of our efforts to become great leaders.

Chapter 3

Driving Innovation across the Enterprise

Creating Value through Test Labs and Sandboxes

One of the benefits of forging close partnerships with start-ups and other technology allies are the lessons they offer in approaching innovation and business. Start-ups, in particular, are working off of a clean slate and have fresh perspectives when it comes to creating imaginative processes and novel approaches to innovation.

CIOs and IT organizations can also draw from the test labs and sandbox environments that are synonymous with many tech start-ups. For the uninitiated, a sandbox is a testing environment where software code or other technologies are isolated from the production environment.

Not only are an increasing number of IT organizations making use of test labs and sandbox environments, they're also gaining value from them. According to a March 2016 study conducted by IDC, 58 percent of IT organizations that are "thriving" have an IT culture that excels at experimentation in every part of the business.

There are numerous benefits that can be gleaned from the deployment of low-cost test labs and sandboxes. These include opportunities for IT team members to gain hands-on experience working on next-generation innovation efforts that have the potential to add meaningful value to the business.

Sample scenarios may include testing an Internet of Things (IoT) device for web interface vulnerabilities before putting the apparatus into production by a utility, or exploring next-generation RFID applications that could be developed for customers in the hospitality industry. IT teams can also use test labs to pilot digital interfaces to help streamline end-to-end business processes.

Test labs and sandboxes also offer opportunities for IT teams to explore new ways for using or applying technologies by the business. For instance, new analytics capabilities continue to emerge, offering end users quantum leaps in scale and speed. Test labs offer IT teams opportunities to test new ways for analyzing data before determining whether they're well-suited for senior executives or other potential business users.

Another benefit of test labs is that they can be used by the IT organization to seek out business problems and opportunities for disrupting markets and catering to unmet customer needs. Case in point: Apple didn't invent the MP3 player but the company did make it substantially easier for users to download music through the creation of the iTunes online music store.

HMG Strategy is working with venture capitalists to help companies develop test labs and sandboxes for innovation. By partnering with industry thought leaders, these efforts are creating new opportunities for IT staffers to play a hands-on, active role in innovation and for the CIO to help the company to grow the business.

"To me, innovation is really about thinking of ways to do things differently," Jean Hill, managing director, Alvarez & Marsal, said in an HMG Strategy video. "Instead of paving over a cow path, how can I fly over the mountain? What is the business outcome that I'm looking to do, and what is the most efficient way, cost-effective way, highest profit-margin way to go about doing something?"

Supporting Incremental Innovation

In the modern enterprise, multiple classes of innovation can emerge based on a company's objectives and the approaches used. Whether it's experiential, sustaining, product, service, process, channel, or other types of innovation, there are numerous avenues that companies can explore to drive improvements in the customer experience and in other areas of the business.

Lately, there's been a tremendous amount of focus around disruptive innovation to create a new market and a new class of customers. Netflix is a classic example of a company that created a new market through its digital streaming services, forcing Blockbuster into bankruptcy.

While disruptive innovation can be extremely attractive and offer incredible business benefits to companies that are successful, incremental innovation provides another path for enterprises to achieve a series of smaller improvements to products, services, operations, and processes.

Still, despite the benefits that incremental innovation can offer to companies, there are a number of risks and challenges that companies must contend with. For instance, the top internal and hidden risks associated with incremental innovation projects include a lack of coordination within the company, lengthy development times, and challenges related to having a risk-averse culture, according to a study of 1,700 projects in 32 European companies conducted by researcher and author Altin Kadareja.

Meanwhile, the most significant external innovation risks associated with incremental projects include uncertain demand for innovative goods or services, lack of customer insight, and excessive perceived economic risks, according to Kadareja's study.

There are a number of ways that CIOs can help drive incremental innovation within their companies while mitigating the risks associated with these projects. As a valued member of the executive team, the CIO can offer expertise in project management to help ensure that the people with the necessary skills are added to the project team and confirm that these efforts are well-coordinated and executed. Proper coordination and management can also help ensure that incremental innovation projects are kept on course and don't become runaway projects.

Because incremental innovation can also be applied to gradual process improvements, the CIO can apply his or her unique view across the enterprise to identify processes that should be eliminated, repaired, or improved upon. Meanwhile, the CIO can also bring technology to bear to help drive process improvement. This can include the use of process mapping tools, which can be used to assess business operations and identify gaps and problem areas that can be addressed.

One of the most important ways that CIOs can add value to incremental innovation efforts is by identifying how and where technology can be applied. For instance, agile development techniques can be applied to mobile apps developed by a company for its customers that can be applied to adjust to changing customer behaviors or interests.

As practitioners know, true innovation doesn't always need to result from a radical change or a disruptive event.

"One of the things I get most excited about is when innovation comes in small, unexpected ways," Jennifer Hartsock, Global CIO of GE Oil & Gas, said in an HMG Strategy Transformational CIO video. "People are waiting for the big bang, but sometimes the most interesting innovation comes from some small nugget idea that then is nurtured and delivers interesting business value," Hartsock added.

Spearheading a Data-Driven Cultural Transformation

Before Jason Cooper joined Horizon Blue Cross Blue Shield of New Jersey as chief analytics officer in February 2015, analytics for the state's largest and oldest health insurer was decentralized, sometimes duplicative, and while quite effective, not always optimally efficient.

The company's analytics-oriented technology infrastructure and human capital investments had fallen a bit behind the times. Jason says, "We weren't deriving as much tactical and strategic value from analytics as we should have been."

Under Jason Cooper's leadership, Horizon BCBS of New Jersey centralized analytics and kicked off a rigorous demand management assessment within its IT, Strategy, and Business Process Improvement (BPI) divisions.

That assessment led Jason and his 100-plus person analytics team to formally launch an analytics transformation project and identify four areas for investment: organizational change management, from how the analytics team was structured, stakeholder management, and workforce planning; analytics infrastructure; self-service analytics via data visualization tools for senior executives and other contributors; and the automation of manual processes and rationalizing low-value work that could be eliminated.

Self-service data visualization tools are now enabling senior executives and other leaders to track the performance of the patient-centered medical home (PCMH) and accountable care organization (ACO), as well as monitoring key performance indicators (KPIs) for divisional units and across the enterprise.

One example of the top benefits realized through the use of the self-service visualization tools has been for root cause analyses behind first-call resolution rates for customer service associates with members. Jason says, "This will eventually help us to better understand which customer service associates are knocking it out of the park (in terms of exceeding member needs) and to identify opportunities for additional training or support that may be needed by other associates."

The success of the company's self-service initiative is hardly surprising to industry experts. According to a 2016 study conducted by Forbes Insights, the most successful business intelligence programs are significantly more likely to place analysis and decision-making tools in the hands of business users.

Upskilling and Reskilling the Analytics Team

Jason has also worked to strengthen the technical skill sets of his analytics team, who already possessed deep business acumen before he came onboard. Targeted training has included SAS, Tableau, R, Python, Hadoop, Spark, and Aster, an advanced analytics platform from Teradata. Team members have also been coached in data visualization and consultative techniques. He has also instituted analyst and leadership user groups that meet to discuss hot topics and knowledge sharing. He added a data science team to complement the company's advanced analytics capabilities.

One of the top initiatives that the team at Horizon BCBS has focused on is the company's value-based transformation—moving the state of healthcare in New Jersey from fee-for-service to a fee-for-value model. As part of these efforts, Horizon BCBS of New Jersey participates in the Omnia Health Alliance, a partnership model with many of the leading delivery systems throughout the Garden State, including Atlantic Health System, RWJBarnabas Health, Hackensack Meridian Health, Hunterdon Health, Atlantic Regional Medical Center, Inspira Health Network, and Summit Medical Group.

Jason says, "In order to succeed in this attainable but challenging goal, we have to provide insights and decision support tools to both our internal decision-makers as well as our healthcare system partners."

Beyond supporting the alliance's triple aim—improving quality, reducing cost, and improving experience—analytics is also being used proactively in the development of predictive and prescriptive models to improve healthcare.

For instance, the analytics team utilizes LANE (Low Acuity, Non-Emergency) algorithms to showcase events where individuals have sought care in the emergency room for clinical episodes that could be cared for in urgent care or primary care facilities.

"Large integrated healthcare organizations are starting to see where they can divert people with ear infections or other medical conditions that can be treated at urgent care facilities," says Jason. Although his team is still in the early stages of quantifying the outcomes for these approaches, the use of the LANE algorithms has shown promise in providing patients with more efficient care more cost effectively, while potentially reducing patient volumes and wait times in hospital emergency rooms.

Meanwhile, the development of opioid addiction models to better understand the statewide opioid epidemic have begun to help Horizon BCBS to target those areas in greatest need of assistance, such as helping to identify the most efficient use of resources.

Through its decision-support initiatives as well as the distribution of self-service data visualization tools to senior executives and other stakeholders, the analytics team provided business leaders the ability to make real and near-real-time, informed decisions based on solid data and insights. Meanwhile, the same insights-driven business decisions—enabled by analytics—are playing out daily in several divisions across the enterprise.

Jason attributes much of the company's success to teamwork. "The progress we've made in the two years I've been here wouldn't have been achieved without a high level of collaboration between our CIO, CTO, and organizational leaders."

Fostering Open Innovation to Drive New Growth Opportunities

Innovation comes from many sources. Employees offer a font of useful ideas for improving internal processes and new products and services. Business partners that work closely with a company can also provide unique perspectives based on their view of the enterprise and industry markets.

Increasingly, companies are reaching outside of their traditional supply chains to collect additional sources for innovation, including customers, external designers and engineers, and other outside contributors. *Open innovation* includes both internal and external ideas, as well as internal and external paths to market.

According to a study conducted by the University of California at Berkeley, roughly 80 percent of enterprise firms in the United States and Europe have open innovation practices.

Potential benefits to the use of an open innovation model include the inclusion of customers early on in the product development process, as well as an expanded pool for ideas.

Still, there are possible downsides to open innovation, including the potential for exposing new product and service opportunities to competitors.

For CIOs who are looking to foster open innovation, the first step is overcoming cultural bias toward internal ideation. A good starting point is by sharing examples of other companies that have jumped ahead of the market and gained competitive advantage through the use of open innovation.

Starbucks has been a poster child for open innovation. Thanks to the creation of its crowdsourcing platform, My Starbucks Idea, the website has received more than 190,000 ideas, with about 300 that have been implemented. These include the deployment of free WiFi at Starbucks outlets, drive-through mobile payment enablement, and free birthday treats for customers.

CIOs can also highlight the business opportunities that are lost when companies fail to grasp open innovation. A study conducted by Accenture finds that enterprises that overlook opportunities for using open innovation are placing $1.5 trillion in growth at risk.

Identifying a respected business leader who will champion open innovation can also hold sway with the C-suite. Small pilot projects that require minimal resources can be used to demonstrate the value of open innovation and serve as a test bed.

For instance, a growing number of companies are making use of crowdservice where a network of customers is helping fellow customers with product or service issues. Crowdservice is a way of offering customers another customer service channel where they can lean on other customers that have encountered and worked through product or service issues.

Within the IT organization, CIOs can also create sandboxes for innovation to experiment with emerging technologies that could potentially be used by the business to deliver new products or services ahead of the market.

While there are numerous benefits to open innovation, the CIO's role is also to ensure that new approaches to innovation make sense for the company.

Fresh Approaches to Kickstarting Innovation

Innovation is no longer an option for companies. Global competition is fierce across all industries. New market entrants and existing players are continually introducing new business models and methods for serving customers that are forcing company leaders to identify and act on new opportunities for improvement and brand differentiation.

The CIO plays a multidimensional role in innovation strategies. He or she is expected to identify how technology can be leveraged to drive new opportunities for the enterprise. CIOs also must draw on their broad view of the enterprise to identify places where innovation can be unleashed.

As a central process coordinator within the enterprise, the CIO is also responsible for recognizing when innovation has stagnated or can otherwise be re-energized to ensure that innovation efforts are firing on all cylinders.

This can start with a workgroup exercise exploring the approaches that the company currently takes to launch innovation efforts and whether there might be new or better ways to approach innovation initiatives. These evaluations may include the number of people or the types of roles that are involved in preliminary brainstorming. For instance, should there be representatives from sales or marketing involved when innovation projects explore new product or service ideas?

CIOs can also closely examine the current checklists that team members adhere to in early and mid-stage innovation efforts and whether any items are unnecessary or could otherwise be streamlined. For instance, could the process for gathering requirements data for an early-stage initiative be approached any differently?

The CIO can also play a critical role in evaluating whether technologies can be used in new or different ways to help unearth new ideas. For example, speech analytics can be used in a company's contact center to not only identify potential support issues that need to be addressed but also to collect, identify, and act on potential product or service ideas that are shared by customers during their interactions with the company.

CIOs can also question whether potential products or services could be tested differently with customers and consumers than the current tactics that are being used.

Concept testing is most often used to test the likely success of a new product before it goes to market. Concept tests that are used to identify the perceptions, preferences, and needs of a product or service sampled by users could be approached differently. For instance, the test team could explore whether the demographic makeup of the test group could be expanded or contracted and what the potential benefits or risks are for either approach.

Since concept tests are conducted in the predesign stage, the test team could also approach concept testing differently than it has in the past. This includes spending more time exploring other potential uses for a product that could lead to greater revenue potential. For instance, before gunpowder was used for fireworks and firearms beginning in the tenth century, the Chinese experimented with it for medicinal purposes, such as a treatment for skin conditions.

As Evan Carstedt, managing director at Accenture, points out in an HMG Strategy video, innovation is about thinking about the business in different ways. "Instead of thinking about just products, how can you think about products and services? You can sell a tire or you can sell a tire with monitoring that creates a whole new business model as well where you can potentially sell a tire as a service. There's a number of ways that businesses are completely reinventing themselves if they understand what business they're truly in."

The CIO as Chief Disruption Officer

Successful CIOs have learned to align IT with business strategy. The same holds true for innovation. Top CIOs pay close attention to how the business defines innovation and then align IT to strategy to match those definitions.

Doing so removes a lot of the hurdles that can arise when CIOs and IT teams are attempting to execute on business-critical IT projects where innovation is involved.

But there's another role that the CIO can play in both fostering and even leading the innovation charge. This is where the CIO can serve as a disruptive innovator.

Disruptive innovation is all around us. We see it on a daily basis with consumer-led services such as Uber and Airbnb. Startups with disruptive business models have emerged in nearly all industries, including financial services (consider Smart Money Capital Management, a computer-assisted financial management company, which invests in exchange-traded funds and charges clients relatively low asset-based fees).

Meanwhile, in healthcare, which continues to undergo a radical transformation, companies such as Butterfly Network have emerged that provide novel approaches to leveraging connected cloud capabilities with diagnostic and therapeutic imaging platforms.

What's equally exciting—and terrifying—about disruptive innovation is that it is open to all players. When we think about disruptive innovators, it's often companies such as iTunes and Twitter. But disruptive innovation can also be driven by established companies. For instance, Apple has an edginess to it, but its biggest innovations such as the iPod and the iPad have all been invented in the last 15 years. Not bad for a company that started in a garage in 1976.

While Netflix hasn't been around nearly as long (founded in 1997), the company reinvented itself after it initially disrupted the retail home movie and video game rental industry with its mail-order business, only to introduce streaming 10 years later.

Because of their view across the organization, CIOs are uniquely positioned to identify opportunities to drive disruptive innovation and digital transformation. But as we discussed before, this requires the CIO to view innovation from the business's perspective.

Consider the rampant use of smartphones and mobile devices by consumers. A recent study of US consumers by Bank of America reveals that 89 percent of Americans check their smartphones at least a few times a day while 36 percent admit they are constantly checking and using their devices.

Think of the opportunities this presents for companies across a range of industries to strengthen engagement with customers and provide the types of mobile services that consumers would want. A CIO for a hospitality or travel company could help identify the potential for creating a mobile app or a mobile website that could make it easier for customers to book a hotel room or a flight, check existing reservations, and drive other transactions using their mobile devices.

CIOs can also draw on their unique view of the enterprise to help identify opportunities for applying disruptive innovation. A CIO for a consumer packaged goods (CPG) company could identify how an emerging technology could be used to gather, analyze, and act on consumer feedback quickly to determine the viability of a new product being market tested faster than its competitors are able to. A CIO for an automotive manufacturer may identify an opportunity to create a new service for customers based on connected car data that's continually transmitted back to the company.

"When we talk about innovation from a CIO's perspective, I think the 'I' in CIO is innovation," says Vic Bhagat, EVP, Enterprise Business Solutions & CIO at EMC in an HMG Strategy video. "We have to figure out how to co-innovate for solutions and even product. We have to solve a business problem. We first have to listen where the business is headed. What are they trying to? Where are they trying to grow? What market are they trying to penetrate? What market are they trying to capture? And how can IT innovate the right solutions to help the business to accelerate—and deliver?"

Learning and Sharing Cultural Nuances of Cybersecurity across Four Continents

Kirsten Davies has had an impressive journey as an information security executive. After starting her career as an independent consultant around IT transformations, she was recruited by Deloitte Australia to help client companies shape transformational work around finance, HR, and IT, understanding and mitigating the risks behind implementing or sunsetting large enterprise systems such as ERP. This early work provided her entry into cybersecurity.

Since 2009, Kirsten has been a veritable globetrotter, having held senior cybersecurity positions at Siemens, Hewlett-Packard/HPE, and currently, at Barclays Africa Group in Cape Town, South Africa, where she is chief security officer.

From her childhood into adult years, her journeys have taken her from the United States and Canada to Australia as well as numerous countries in Europe before joining Barclays Africa Group in February 2017. To help put this in perspective, she flew more than 200,000 miles last year, spending the equivalent of three months in the air.

Throughout her travels, Kirsten has learned invaluable cultural nuances with respect to cybersecurity and has also been able to apply some of these across her various stops. "It's a constant engagement mechanism to build awareness, understanding, and partnerships. We don't own what we have to secure, such as the IT infrastructure. So you're constantly building awareness and partnerships to execute."

She is also reminded that people are people wherever you go and not to make assumptions as to where they are on their own respective awareness journeys in information security.

"When engaging on subjects of security with CEOs or board members, that's the first lesson I've learned. The second lesson is that you can't presume where the organization is, either, in its maturity level for IT or for security. You can come in and be a rock star but if the IT is outsourced, there are multiple layers of legacy IT at varying degrees of process maturity, or if the enterprise has a very basic or nonexistent security culture, you're starting from the drawing board."

Meanwhile, the more companies that Kirsten has worked for, the more she realizes that CISOs have one of the toughest jobs there is. "It's a core function, not a business unit and not just IT. We can often be treated like we're not core, like a blocker, but we have to get things right—the partnerships, the execution, the board visibility and support—because if there's an event, the impact on the enterprise can be huge."

There's also more need than ever for CISOs to have a high EQ, or emotional intelligence quotient, to help partner in enterprise risk strategy. "We've been the quiet protectors behind the keyboards and buried away to scan and fix the code," says Kirsten. "But security is not an IT problem—it's an enterprise risk problem—and we have to increasingly be on the front foot, in the middle of strategy discussions, because our mandate is at the center of organizational strategy."

Opportunities to Add New Tools to the Toolbox

Kirsten sees the experiences she gleans from each of her information security roles as an opportunity to add new tools to her toolbox.

"As a professional and as an executive, I would think that everyone has the same mindset that the last opportunity builds to the next opportunity you have," she says. "Having worked in so many different countries, I've been able to add new tools to my toolbox and I apply them each day in my role. Some of the tools get worn out and have to be discarded. But with other things, you can say, 'I picked up this tool in Germany and I can use this tool in South Africa.' That's very much part of the day-to-day in my job. I do believe wholeheartedly that my work across the world has helped me to step into my new role."

Chapter 4

Leaders of Transformational Change

As Connected Global Economy Grows, CIOs Rise to Meet the Challenge

I've been writing about the elevation of the CIO for more than 10 years, long enough to see what began as a hopeful vision become a firm reality. CIOs have earned their seats at the table and become trusted partners with their peers in the C-suite.

Building on that foundation of trust, many CIOs have moved higher in the enterprise value chain, evolving into strategic business leaders and key drivers of long-term growth.

Scott Fenton, VP and CIO at Wind River Systems, is an excellent example of today's rising technology executive. Wind River is a world leader in embedded software, an area that has become a sweet spot for growth in the connected global economy.

Wind River was acquired by Intel in 2009 and is considered an essential technology vendor in the rapidly expanding markets created by the Internet of Things (IoT). The McKinsey Global Institute estimates the IoT will generate somewhere between $3.4 and $11.1 trillion annually in economic value by 2025, and the World Economic Forum (WEF) estimates it will generate $14.2 trillion by 2030.

As the IoT gains traction as a real business opportunity for thousands of companies, premier technology providers such as Wind River are perfectly positioned for success.

"This is a very exciting time to be at Wind River," says Scott. "We're right in the middle of the IoT revolution, which is great. We've been providing embedded software for 30 years, and our experience can't be matched."

Embedded software will be absolutely fundamental to IoT growth. That said, embedded software is difficult to create, and many developers are caught off-guard by its complexities. A big part of the challenge is finding developers who genuinely understand the complicated relationships between software, hardware, manufacturing, and supply chain optimization.

Wind River's track record gives it a unique advantage in the growing IoT market, and also elevates Scott's role as an executive leader. "My role and responsibilities as CIO have evolved over the past two years," says Scott. "In the past, CIOs were mostly responsible for running internal systems. Today, CIOs are more focused on growing business revenue and helping their companies become more successful. For me, this is incredibly exciting, and it's a great time to be a CIO."

I'm looking forward to hearing more from Scott and other outward-facing CIOs as they continue their upward trajectories at this fascinating moment in our history.

Detroit Sees Its Future in Digital, Delivering Superior Customer Experiences

Detroit is an amazing city. For decades, its economic power and vitality were emblematic of American culture. Then, for a variety of reasons, Detroit lost its luster. Many wondered if the Motor City would ever recover.

Today, Detroit is experiencing an economic renaissance, thanks largely to the powerful resurgence of the automotive industry. We hold an excellent series of CIO Leadership Summits in Detroit, and one of our featured speakers in 2017 was Raj Singh, EVP and CIO at FordDirect.

I had an absolutely fascinating conversation with Raj and I was impressed by his ability to blend superior business knowledge and expert digital technology skills to create real value for one of the world's leading brands.

FordDirect provides digital marketing and advertising solutions to Ford and Lincoln dealers, giving them a uniquely valuable platform for driving more sales. For example, FordDirect provides an array of dealer services such as marketing solutions, websites, leads, and e-tools that dealers can use to improve sales, service, and customer satisfaction.

With a foundation built by Ford Motor Company and its franchise dealers, and being the only joint venture of its kind, FordDirect genuinely understands the automotive and dealer business.

"Five or six years ago, the conversation would have been about finding an IT solution for the supply chain. Today, the conversation is about the customer experience," says Raj. "We're working on making each customer's experience faster, better, and smoother so they can make their purchase decisions more quickly. Today, customers complete more of the car buying transaction online. You want to trade a car, you want a replace a car, you want to lease a car, you want to finance a car—whatever you want to do, you can complete all of it online, in just a few minutes. That's amazing."

Raj is definitely a high-energy type of leader, and that's exactly what the auto industry needs at this critical moment in its history. I'm betting that Detroit's rebound will continue. According to my sources in the community, Detroit has become a magnet for some of the best and brightest developers and designers in the world. As most of you know, the top auto makers have already launched their own software startups to compete with Silicon Valley.

Imagine Detroit competing with the Bay Area for top tech talent. That certainly would be an interesting battle.

IoT Governance Challenge: Looking for "One Throat to Choke"

The Internet of Things is expanding, and soon it will be everywhere. By 2020, we'll have billions of connected devices, including drones, cars, robots, cameras, thermostats, dishwashers, and refrigerators.

The technologies behind the IoT are evolving rapidly. But the legal and regulatory pieces of the puzzle are still emerging. Who do you call—or sue—when something goes wrong?

Those aren't idle questions. Finding answers to those types of questions will be essential to spur investment in IoT companies. For example, let's say the IoT-controlled smart elevator in your dentist's office breaks down and you're stuck in it for three hours. Who gets the bill for your lost time and aggravation? Who is responsible?

With the IoT, it will be hard to assign direct responsibility. Many entities will be involved: the building's owner, the elevator's manufacturer, the company that maintains the elevator, the phone company that provides the network linking the elevator to the IoT, the companies that made the sensors in the elevator and probably even your dentist.

Chances are, you'll be looking for "one throat to choke," as attorneys are fond of saying. But with the IoT, there will be numerous throats—far too many to hold accountable.

And it will only get more complicated. Technically, driverless cars are part of the IoT. But in an average day, your driverless car might drive across streets, roads, highways, and bridges that are built and maintained by hundreds of separate municipalities and jurisdictions. Which one is responsible if your car loses its network connection and hits a guardrail?

Invariably, CIOs and corporate IT departments will become enmeshed in problems caused by or related to the IoT. It's not too early to begin thinking about IoT governance and guidelines. Undoubtedly, many corporate assets will be connected to the IoT, and it's only a matter of time before IT gets involved.

In all likelihood, IoT governance will become part of IT governance. Unlike IT governance, however, IoT governance will extend far beyond the traditional walls and boundaries of the modern enterprise. Again, the role of the CIO will expand to take on the new responsibilities spawned by a new world of connected devices, sensors, and machines.

IT Leaders Create Tangible Business Value and Deliver World-Class Customer Experiences

My good friend Ralph Loura is CTO at Rodan + Fields, a leading prestige skincare brand and social commerce company.We had a great conversation recently about the evolving role of the IT leader as a value creator for the enterprise, and I want to share some of Ralph's insights with you.

"IT leaders play meaningful roles in virtually every company and in every industry," says Ralph. "This is a very cool time to be an IT leader. We've moved from the back office to the front lines. We're full-fledged partners with the business and the expectations are higher than ever before."

In the past, IT leaders were often constrained by high costs and long development timelines. Today, the cost of deploying new systems has fallen dramatically. Thanks to the cloud, it only takes days—or in some case, even hours—to deploy new IT capabilities.

"Today, when the business needs to create a web-based storefront, the IT team can set it up easily and cost-effectively with a cloud-based service. You couldn't do that five or six years ago," he says. The cloud creates incredible opportunities for IT leaders to become real business leaders in the modern corporate environment.

"As a group, IT leaders are maturing into seasoned executives with strategic vision," says Ralph. "We've risen through the ranks and we're ready for leadership roles."

IT executives are also stepping away from their traditional roles as service providers, and embracing how customer-centricity drives business success in competitive markets.

"All IT leaders and executives need to embrace models that are user-centric," says Ralph. "Our role is enabling the business to provide amazing experiences to our customers.

When you look at great companies today, they all focus on the needs of their customers and users."

Ralph cites Uber as an example of a business that's grown impressively by focusing diligently on the needs of its users—people who want to get from Point A to Point B with the minimum hassle. You could say that Uber is a taxi company or that it's a technology company, but both of those descriptions would miss the point. Uber is a personal transportation company, and that's the key to its success.

"In the early days of my career, I had the privilege of working with some great CIOs," says Ralph. "They didn't see themselves as ordertakers. They understood the business and they were involved in developing business strategy. They worked with external customers, and they played important roles as executives guiding the business toward greater success."

I genuinely respect and admire Ralph's deep understanding of the new IT leadership role as a trusted value creator and business partner. As Ralph often says, it's a great time to be an IT leader, and I totally agree with him!

Courageous Leadership Builds on a Solid Foundation of Skill, Experience, and Trust

I had an excellent conversation with my friend Guy Chiarello, the president of First Data Corporation. Before joining First Data in 2013, Guy was the CIO at JPMorgan Chase, where he was responsible for technology across seven lines of business in more than 60 countries. In his role as CIO, he oversaw a vast infrastructure and drove innovation, helping the bank launch a widely successful suite of digital banking platforms, including the award-winning Chase Mobile App Suite, which grew its customer base to more than 10 million users in just a couple years.

I mention Guy's amazing credentials as an executive leader because our conversation focused primarily on leadership and the critical importance of developing deep leadership skills over the course of a successful career.

"I've worked with a lot of very smart people, and I always learned from them," says Guy. "In many leadership roles, you cannot take the middle ground. You have to act boldly and reach for things beyond your grasp. You need the confidence that comes from knowledge and experience."

Very early in his career, Guy learned that successful leaders always think in terms of business or commercial impact. "Especially if you're at a Fortune 50 or Fortune 100 company, you've got to be thinking about the commercial potential of your projects. That's the only way to drive transformational change," he says. "Always look beyond the technology and focus on the business impact."

In addition to understanding the business, successful leaders are courageous. "You need the ability to see around corners. You won't always be right, but you need to keep moving forward. Don't let yourself get buried in the technology."

Guy and I spoke about the difference between "swagger" and "courage." We agreed that courageous leadership is built on a foundation of competence, confidence, performance, solid business relationships, and trust. You can't fake it or adopt it as a "management style." For leaders, true courage develops over time, and through experience. In other words, you have to earn it.

"To a large degree, successful leadership depends on good, old-fashioned hard work," says Guy. "You want to be the absolute best at what you do, but you also need to keep moving forward. Great leaders step back and schedule their priorities; they're always learning and they're always getting better. You can never be perfect, but you can always keep learning."

Are You Catching Up to the Cloud or Leading the Way?

My second business book, *On Top of the Cloud,* was published in 2012. It seems like a long time ago. So much has changed. Today, we take social media seriously. There's a mobile app for everything. We're all trying to figure out how to leverage advanced analytics and extract value from big data. Everyone is asking us if we're prepared for the Internet of Things.

It seems odd, however, to hear senior IT leaders debating the merits of cloud computing in 2017. Frankly, I thought the debate was over. But I still hear people talking about the cloud as if it were some kind of passing fad or flavor of the month.

As I wrote in that book, the cloud is a lever for transformation. It is disruptive technology *par excellence,* and it cannot be ignored. The cloud is here to stay—it's a permanent part of the modern IT landscape.

Let's face facts: The cloud is the fastest, easiest, and most cost-effective way for spinning up additional computing power when you need it. And the best part is that when your need diminishes or goes away, you can stop paying for it. You can't do that with heavy iron computing infrastructure. Imagine calling up one of the big mainframe vendors, telling them you're done using their equipment, and asking them to take it back until you need it again.

Instead of fighting the cloud, we should focus on getting ahead of it. I don't have any intrinsic problems with "shadow IT," but we shouldn't let it define enterprise cloud strategy. Then we'd have to play catch up, and that's not an ideal strategy.

The argument I hear most frequently is the cloud isn't secure. But the truth is that the major cloud providers have significantly more experience managing and assuring security than most IT departments. From my perspective, your data are probably safer in the cloud than in your data center.

That doesn't mean I'm recommending that you outsource your security to a cloud provider. But most cloud providers provide excellent security. They have to provide the best security, or they would be out of business.

Rather than becoming cloud administrators, CIOs should become cloud architects. Instead of merely following "shadow IT" into the cloud, we should provide leadership, vision, and strategy. It's always better to lead than it is to catch up.

At CarMax, Technology Drives Great Business Outcomes with Transparency and Customer Focus

I had an absolutely fantastic conversation recently with Shamim Mohammad, SVP and CIO at CarMax, the largest used car retailer in the United States. From my perspective, CarMax is the perfect example of a great company built on a rock-solid foundation of superb information technology expertise and execution. CarMax is a Fortune 500 company, and has been named on FORTUNE magazine's 100 Best Companies to Work For® list 13 consecutive years.

At CarMax, Shamim is part of the company's senior leadership team, and that means IT's voice is always present at the highest level of the firm. "We meet every Monday, travel together, and work together as a team," says Shamim.

"The whole leadership team is committed to the ways in which information technology plays a critical role in the customer experience. It's important for the CIO to help lead the conversation on how to make every aspect of the customer experience better in our data-rich environment. With customer expectations changing rapidly, I cannot be successful as the CIO unless I also consider my role to be the chief innovation officer and chief idea officer," he says.

At CarMax, customers are in the drivers' seat, which explains the company's phenomenal success. Since being founded more than 20 years ago, CarMax has transformed the way used car buying is done by providing the honest and transparent experience customers deserve.

"We were the original disruptor of the industry, offering a no-haggle and no-hassle experience, and we continue to revolutionize car buying through customer-focused technology innovations," Shamim says. "Nine of 10 of our purchasers start their car search online, and technology is a critical part of our mission to deliver an unparalleled customer experience. Part of our digital transformation has centered on our evolution to an omni-channel retailer. We're committed to delivering a seamless experience for the customer throughout their journey from digital e-commerce to in-store, and will be there for customers however and whenever they want to shop."

It takes ultra-high levels of precision and accuracy to execute consistently on the firm's business model. For CarMax, great IT isn't a luxury—it's absolutely foundational.

"Enabling our associates with the right tools and analytics is critical. Our product teams are structured into small, agile teams of cross-functional and omni-channel associates that are empowered to go after key results rather than being told *how to* solve a problem. This Silicon Valley—based model allows our associates to iterate quickly with autonomy to design for the end user in mind," says Shamim.

I genuinely respect and admire how CarMax has built a world-class brand by focusing on the customer first and leveraging technology to compete successfully in a highly turbulent market.

"We know that CarMax isn't being compared to the auto dealership next door, it's being compared to the Amazons and Starbucks of the retail industry. Innovative technology allows us to deliver a personalized experience that customers expect."

The CarMax story also demonstrates the value of learning lessons from related industries and applying those lessons in disciplined ways. The company's approach to price transparency is similar to the approach taken by the retail industry. Retail consumers are generally aware of prices before entering a store, and surprises are kept to an absolute minimum. In the past, many used-car dealers relied on opaque pricing strategies to confuse consumers. The success of CarMax, on the other hand, is based on providing total transparency to consumers.

"We're committed to integrity and a culture of respect and transparency," says Shamim. "We continue to focus on how we can make our customer and associate experience even better through customer-facing technologies, and we're always looking forward to see what's next."

Transformative Change Is a Universal Phenomenon

There's a recent adage that all companies are now technology companies due to their reliance on technology to run the business and use data to drive decision-making.

Taking that logic a step further, Monsanto CIO Jim Swanson points out that all companies are undergoing some type of transformational change right now, whether it's connected to M&A, R&D, business disruption, or some other factors.

For its part, Monsanto is undergoing transformative change on at least two fronts. In addition to its announced merger with Bayer AG, Monsanto is digitizing many aspects of its agricultural business.

For instance, Monsanto is gathering and analyzing real-time information from combines that are used on farms to determine how they are executing. The company is using digital tools and data to ascertain how its agricultural products are performing.

In short, Monsanto is leveraging data and digital technologies to bring farming into the twenty-first century. One thing that Swanson has learned from his experiences with digital disruption is that time compression is placing even more pressure on executives to execute.

Jim says, "As companies try to figure out how to navigate through disruptive change, they need to have iterative approaches that deliver business capabilities in weeks." He spoke about Monsanto's digital strategy at HMG Strategy's 2016 St. Louis CIO Executive Leadership Summit. "Companies that don't move fast enough aren't going to stick around. As CIOs and technologists, we have to help the CEO and the board to deliver value. It requires courageous leadership."

One of the ways that Jim and his team are doing this at Monsanto is by being proactive—and not reactive—with its digital initiatives. This includes the company's endeavors to monetize its digital efforts.

"You have to be very forward-looking in creating a market, not responding to it," he says. "By waiting on a market, we lose our ability to shape the future." Meanwhile, it's not solely up to the business to move aggressively on digital. "Every single role (in IT) needs to think about how to disrupt the business."

Still, he believes it's not enough to merely generate some level of digital disruption. "If we're not disrupting by a factor of 10x, we shouldn't think about doing it."

To help communicate the value that digital initiatives are delivering to Monsanto, Jim's team uses video and other types of messaging to spread the word across the enterprise.

The crucial role that Monsanto's IT organization is playing in the company's digital transformation has stoked Jim Swanson. "There's no better time to be a CIO and to drive change."

Building Strong Bridges across the Enterprise with Humanity and Humility

I had a great conversation recently with Joe Topinka, the CIO of SnapAV and the author of IT Business Partnerships: A Field Guide. Joe credits much of his success as a CIO to always remembering that it's important to treat colleagues and customers as human beings.

That might seem like an obvious piece of advice, but in today's fast-paced business environments, we sometimes forget that most of our interactions are with people. Most of the people we encounter in the course of a typical day have hopes, dreams, goals, and aspirations that are very similar to ours.

It's especially important advice for CIOs as they build relationships with other executives in the C-suite. A main part of the CIO's role is bridging the chasm between IT and the rest of the enterprise. Building solid relationships with colleagues across the enterprise requires more than great technical skills—you need healthy doses of humility and humanity to convince people you can be trusted over the long haul.

Joe has developed a useful list of nine rules that apply to work and life. Here's his list:

1. Assume positive intent.

2. Say "Yes and...."

3. Be respectful.

4. Put facts before stories.

5. Listen first.

6. Be present and engaged.

7. Support it—don't debate it.

8. Be accountable.

9. No BMW (bitching, moaning, whining) driving.

"Once you get your head around personal accountability and what it means, everything else becomes easy," says Joe. "The frustration goes away and is replaced with positive energy that leads to positive outcomes. I know it sounds a little philosophical, but for me, it's been game changing. I wish I'd figured this out way earlier in my career."

CIOs should make a point of meeting personally with external customers and should be fully capable of articulating the value of products and services, says Joe.

"Bridging the chasm requires speaking the language of business. Too often, our inability to speak in business terms hurts us. Look at your portfolio of IT projects and map those projects to your company's business strategy," he says.

"We're responsible for reaching across the aisle. We need to start talking like business leaders."

Joe's advice and insight rings true. It also aligns perfectly with my vision of the CIO rising to become the CEO of IT. From my perspective, that's the real challenge and the real opportunity: elevating the role of CIO from technology leader to enterprise business leader. I'm supremely confident we can turn that vision into reality.

Are Your Enterprise Apps Ready for the Cloud?

There's been a lot written about moving big data analytics and applications into the cloud. The basic concept makes sense for two key reasons:

1. Advanced processes such as artificial intelligence, machine learning, and neural networks run on big data, which often resides in the cloud.

2. Renting cloud instances is usually less expensive than buying custom hardware to run advanced analytics.

But if you're the CIO of a large enterprise, there's a good chance your in-house data scientists have already developed a bunch of home-grown analytics for dealing with specific challenges that arise in various business units across the enterprise.

If those home-grown apps are popular and effective, your CEO will probably ask you about scaling them for the rest of the enterprise. That's when they can become headaches for IT.

An application or solution that's written for enterprise architecture might not work as expected in the cloud. Enterprise solutions are written with vertical integration in mind. In other words, they have to work smoothly with the software running immediately above and below them.

Cloud-based analytics, on the other hand, usually require a high degree of horizontal integration, because they're constantly adjusting to huge volumes of fresh data from multiple sources.

It takes a different mindset to write a cloud-based app. You can't simply take an app that's designed to run on enterprise hardware and transport it magically into the cloud.

Explaining the differences between on-premise enterprise architecture and cloud-based architecture can be difficult, especially if the audience isn't particularly interested in the technical details and just wants to see quick results. But you can't simply pretend that it's easy to scale a local solution into an enterprise solution by merely shifting it into the cloud.

Most executives don't understand why an app designed to run on local equipment won't necessarily run in the cloud. My advice is to begin talking about scale issues now, so they don't become surprises later on.

Take the time to develop a short presentation or video explaining the differences between enterprise architecture and cloud-based architecture.

Begin the dialogue now, and you will avoid a lot of headaches down the road.

The Big Pivot: How Apple is Rocking Our World

Last year, we watched Tim Cook unveil the new iPhone 7. From outward appearances, the new phone represented a modest step forward in tech development. Most of the attention centered on the absence of a traditional headphone socket.

But let's take a deeper look at the phone and what it means in the broader economy. It seems clear to me that Apple is sharing its vision of a wireless future. Throw away your cables and your connectors, says Apple. From now on, connectivity will be a largely wireless phenomenon.

I actually believe this is just the beginning of an amazing transformation. Frankly, I do not believe that Apple and other tech makers will be satisfied by merely creating a wireless world. The real future extends beyond far beyond going wireless.

Remember when Intel pivoted from memory chips to microprocessors and when IBM pivoted from mainframes to services? Those are classic examples of pivots. A more recent example of a world-class pivot would be Tiny Speck's switch from gaming to collaboration, which led to the creation of Slack, the incredibly popular collaboration platform.

From my perspective, Apple is preparing for its own massive pivot, from hardware and devices to cloud-based services. Here's why this makes sense: After you've eliminated the need for wires between devices, the next logical step is eliminating the devices themselves.

When Apple can deliver the features and functionality of an iPhone without the iPhone itself, that's when the real revolution begins.

Make no mistake, we're entering a new era of technology in which hardware and devices will be "behind the curtain"—largely invisible and hidden from view. Virtually all of the data and processing capabilities we need will be in the cloud. Any device we carry will be far smaller and far more powerful than any tablet or smart phone that we carry today.

Very soon, all the computational resources we need will be around us. We'll live in a state of continuous digital connectivity and seamless ambient computing. We won't need to actually carry devices because we'll be surrounded by smart infrastructure.

That's the underlying message of the iPhone 7 launch. The world is changing, and wireless headphones are just the beginning.

Soon, BYOD Will Include Cars

Some of us are old enough to remember when it wasn't okay to bring your own device to work. In the recent past, people have argued vigorously over laptops, mobile phones, tablets, and flash drives.

The next big argument will be over cars. As we all know, the modern car is rapidly evolving into a powerful computer on wheels. Fairly soon, your connected car will be just another digital device.

For many CIOs, connected cars will pose various integration challenges. There will be complex issues to manage, such as security, reliability, and even safety. Like it or not, however, cars will become part of the IT portfolio.

As cars become more autonomous, we'll spend more time working in our cars and less time actually driving them. We will treat our cars as extensions of our workspaces. We will expect the same user experience from our digital information systems whether we are at home, in the office or in the car.

Based on my experience, it seems likely that companies will hold their CIOs responsible for the quality of the user experience, no matter where it takes place. CIOs will have to become champions of enhanced connectivity and network security. We'll have to weigh into the ongoing debates over wireless and cellular standards. We'll need to make absolutely certain that vendor consortiums and regulators consider the needs of enterprise users.

We're at the cusp of the greatest transformation of society since the advent of computers. Google, Tesla, Baidu, Ford, GM, and Uber plan to compete in the driverless car field. Boston Consulting Group estimates that driverless cars could generate $42 billion in annual revenues by 2025. That's a prize worth competing for, and frankly, I think that number is on the low side. And it doesn't really take into account all the other economic consequences of the shift from traditional cars to genuinely smart cars.

Nobody is quite sure where the driverless car economy is heading, but one thing is certain: CIOs will be held accountable for integrating tomorrow's cars into tomorrow's enterprise information systems.

What Tech Execs Can Learn from Google's Driving Lessons

We've all heard and read so much about self-driving cars for the past couple of years that it didn't seem like major news when Google announced recently that it had decided to convert its driverless car laboratory into a real business.

From my perspective, the lack of fanfare actually shows us how rapidly the whole idea of driverless cars has evolved from a brilliant fantasy into a genuine business category.

There's a message here for technology executives: Some ideas can take off much more quickly than expected. We experienced a similar surprise when mobile phones became essential tools for the modern enterprise. In the early 1990s, several IT executives told me it would be impossible to integrate digital mobile phones into their enterprise communications systems.

In retrospect, they couldn't see the writing on the wall. When the C-suite demanded integrated mobile telephony, the IT execs had to deliver. The same thing happened with mobile tablets. At first, IT execs were reluctant to accept the tablets as legitimate workplace tools. But when the sales force demanded them, IT had to figure out how to integrate the tablets.

Driverless cars, and the new technologies supporting them, will pose similar challenges to IT teams. I strongly urge IT leaders to begin thinking now about integrating driverless technologies with enterprise systems.

History teaches us that new technology arrives with astonishing force and speed. Don't get caught off guard; start planning today.

I also urge you to read Alex Davies's excellent story about Steve Mahan's first ride in one of Google's self-driving pod cars. Mahan, whose vision was destroyed by a rare genetic illness, said the experience made him feel like "a whole person again."

Mahan took his history-making ride through the streets of Austin, Texas, which is now calling itself "the Kitty Hawk of driverless cars," an homage to the site of the Wright Brothers' famous first flight back in 1903.

Note to the C-Suite: IT Glitches and Outages Damage a Company's Reputation and Bottom Line

July 8, 2015, was a bad day for the IT industry. The New York Stock Exchange was down for nearly four hours. There were also embarrassing outages at United Airlines and the *Wall Street Journal*. Many commentators wondered if the glitches were somehow related to the sharp slide of the Chinese stock market.

I spoke that day with my friend Shawn Henry, president of the Services Division at CrowdStrike, a globally respected cybersecurity firm. At CrowdStrike, Shawn leads teams of responders who investigate computer network breaches primarily orchestrated through targeted attacks. Before joining CrowdStrike, he was executive assistant director at the FBI and he's considered one of the best minds in the cyber security industry today.

From Shawn's vantage point, the outages were the result of technical issues and not the work of malicious adversaries. The high-profile glitches were most likely caused by network problems and, in one instance, a software update.

"These kinds of issues occur frequently," says Shawn. "That's why you need to prepare for them and practice your recovery procedures. You need a continuity of operations plan, and you need to train your people to execute on it when systems or networks go down. Remember, the goal is getting the business back up and running as quickly as possible."

Whether the outages are accidental or the result of nefarious activities, they should be a source of concern to C-suite executives worldwide. When a company's website goes down or when its customer database is hacked, the public holds the company accountable. The public doesn't really care about the technical details. And once the damage is done, it can take a long time for a company to repair or rebuild its reputation.

That's why the C-suite needs to pay attention—because outages and glitches pose very real business strategic risks, no matter how they happen and no matter who is ultimately responsible.

Can Big Blue Reclaim a Leadership Role among the World's Titans of Technology?

Depending on whom you believe, IBM is either working out the kinks in its global strategy or flailing around in search of a strategy. I'm sure that most investors would agree that no matter what the reason, they'd rather see IBM get its house in order and return to a leadership role in the tech industry.

On a bright note, IBM's partnership with Apple has the potential to be a winning deal for both companies. As Apple describes it, the partnership provides business with a "new class of apps—entirely reimagined for the mobile enterprise made for iOS, and designed to empower employees wherever their work takes them." In the words of IBM, the truly historic partnership with Apple combines "the power of enterprise data and analytics with an elegant user experience, to fundamentally redefine how enterprises empower their professionals to interact, learn, connect, and perform."

From my perspective, IBM's collaboration with Apple is a great idea, with plenty of upside for all parties involved. The partnership also might help IBM overcome negative perceptions about its strategic commitment to maintaining healthy customer relationships across the board. It might even drive up IBM's Net Promoter Score, which is currently pegged at 27 by NPS Benchmarks. By comparison, fast-growing companies often have Net Promoter Scores between 50 and 80. A recent IBM white paper even mentions "a strong correlation between high NPS scores and increased revenues and customer loyalty," so I have to assume that IBM is aware of its shortcomings in that area.

There are definitely signs that IBM is heading in the right direction. The company's "strategic initiatives," such as cloud computing and analytics, are growing, and its mainframe business remains strong.

For most of the twentieth century, the world counted on IBM for technology leadership and innovation. But with the exception of Watson, which is a genuine marvel of cognitive computing ingenuity, IBM has done little to merit praise or admiration in the past five years.

Today, the tech industry looks to Silicon Valley for inspiration and ideas. It would be tragic if IBM followed the lead of other formerly great technology companies such as Xerox and Kodak. Hopefully, that won't happen. IBM has one of the smartest research teams on the planet, and with the right executive leadership team, it could mount a turnaround and reclaim its position as a titan of tech.

It wouldn't be the first time that IBM turned itself around and recovered its bearings in a turbulent market. Back in the 1990s, the legendary Lou Gerstner led an executive team that restored Big Blue to its role as an industry leader. The big question now is whether Ginni Rometty can engineer a similar recovery and lead the company back to center stage. Let's hope that she does.

Chapter 5

Elevation and Career Ascent

Contributing Value to the Board

As business continues to become technology-driven, CIOs and IT executives are increasingly expanding their roles beyond the C-suite.

Their experience and expertise is sorely needed at the board level. According to a 2016 study conducted by Accenture, just 10 percent of board members on corporate boards have professional technology experience.

There's tremendous value that CIOs can offer to boards by illuminating them on critical technology trends couched in business terms.

"Board-ready CIOs are business leaders first with the strongest technology backbones possible," says Adriana Karaboutis, now former EVP Technology, Business Solutions & Corporate Affairs at Biogen. Adriana is on the board of Advance Auto Parts and Perrigo plc, both public companies, and Biogen and Blue Cross/Blue Shield of Massachusetts (not-for-profit). She recently joined National Grid pc as chief information and digital officer.

She says the first thing that boards look for from CIOs is the ability to speak clearly about technology issues in business terms. This includes the capacity to explain the advantages and benefits of different classes of digital technologies, such as social, mobile, analytics, cloud, and IoT, in understandable terms. Equally, CIOs on boards help delineate the most critical topic and risk consuming boards today: cybersecurity.

Given the escalating risks posed by cybersecurity attacks, boards also expect CIOs to keep them apprised of emerging cybersecurity challenges and risk mitigation strategies. One way of doing this is by communicating to the board examples of cyberattacks that have been launched against other companies recently, along with the impact that these attacks have had on operational uptime, margins, customer loyalty, and so on.

Another way that CIOs can demonstrate their value to the board is by communicating the returns generated by investing in data and analytics. This can include educating the board about how the use of market and customer data with ana-lytics can be used to unearth insights into the factors that are contributing to recent changes in profit margins or customer satisfaction, along with actions that have been taken to address specific business challenges and the results that have been achieved.

In working with the C-suite and the board, it's critical for CIOs to be effective communicators and great storytellers in order to effectively express what they're trying to impart.

"I believe future CIOs and future leaders have to be technologists with technical depth and breadth that have deep business domain so they understand the business that they're in, and they can convey the art of the possible," says Snehal Antani, CTO at Splunk, in an HMG Strategy video. "This means that they're great storytellers, and they can tell stories to the board which may not necessarily be tech-savvy."

The New Leadership Mandate in the Digital Economy

Digital transformation is sweeping across the business landscape. Billions of devices are connected to the Internet of Things, providing companies fresh insights regarding customer behaviors and operational conditions.

Meanwhile, the expanded use of digital channels such as chat, mobile, and social by consumers is changing how companies are doing business with customers and leading to the creation of new business models.

According to an October 2015 study published by Forrester Consulting and Accenture Interactive involving in-depth surveys with 396 business decision-makers, the top three drivers behind digital transformation are profitability (58 percent), speed to market (51 percent), and efforts to improve customer satisfaction (48 percent).

Despite the colossal opportunities for driving business value and operational improvement offered by digital transformation, there is still confusion regarding who should "own" and drive digital strategies within the enterprise. According to the Forrester/Accenture study, digital ownership is currently divided between the CEO (38 percent), the CIO (33 percent), and other senior leaders such as the chief digital officer and chief marketing officer (29 percent).

As digital disruption permeates the enterprise, CIOs increasingly must display courageous leadership in an environment where there is less guidance. In light of this ambiguity, there are a number of ways that CIOs can step up and deliver the kind of bold leadership that's needed for the enterprise to execute on its digital strategies.

A good starting point for CIOs is by working closely with the CEO and members of the executive team to identify and prioritize opportunities and challenges related to digital transformation and strategic planning.

One of the ways that CIOs can demonstrate leadership here is by sharing their insights as to areas of the business or specific operational activities that could benefit from digital deployment based on their comprehensive view of the enterprise and how all of the various business lines and organizational functions are connected and interoperate.

Because the CIO has such a unique vantage point across the enterprise, it's imperative for IT leaders to communicate what they're seeing along with recommendations for executing on opportunities that have been identified. For instance, based on their knowledge of the business and understanding of current market conditions, a CIO for a healthcare provider could identify how the use of wearable medical monitors worn by patients could be used to create new information services for both patients and physicians, or how the use of 3D printing can be used to design and develop customized medical devices.

Another way that the CIO can demonstrate digital leadership is by working closely with third-party business partners that can offer useful guidance and expert resources to companies drawn from extensive experience working on digital business issues with organizations across industries. In this regard, the CIO can act as chief partnership officer in helping the enterprise to identify those business/technology partners that offer the best fit for the company's digital vision.

At a time when executive leadership is needed more than ever, there are multiple opportunities for the CIO to step up and deliver value for digital transformation initiatives.

"When I first started in energy distribution, I did find that there was a lot of inertia," says Maurizio Laudisa, CIO at Superior Propane, in an HMG Strategy video. "It's an old industry—it's essentially a lot of people with trucks going and filling up propane tanks. It was really ripe for new technology to break through for the commoditization of that industry. The way we have shaken the status quo was to really re-think how digitization can help the industry look beyond where it's at and disrupt the playing field. We've done that in a couple of ways. We've actually built customer portals that are integrated directly into our supply chains for both our residential customers and our commercial customers."

Leading by Example

The CIO wears a number of hats in the enterprise. The CIO helps to identify and enable opportunities to grow the business and improve the efficiency of the organization through the use of technology. The CIO also oversees the IT infrastructure that's needed to run the business on a day-to-day basis. In addition, the CIO helps to identify productivity tools that can help executives, managers, and employees to do their jobs better and more easily.

As the head of the IT organization, CIOs also set an example for the rest of the group. CIOs aren't leaders because of their title but based on the actions that they take. This includes how they handle decision-making, whether they act decisively, listen to, and act on input from managers and staff, and how they communicate the actions that have been taken.

CIOs show their leadership chops when tough decisions have to be made. This may include how a firing or a lay-off is handled and how it is communicated to the affected individuals as well as the team.

Effective leaders listen to what others have to say and carefully consider their opinions. They consider different perspectives to each problem and they enthusiastically promote great ideas from other people. Great leaders are also highly trusted. Trust has to be earned, and that takes time.

Leaders gain trust in a number of ways. They're clear in their messaging about the organization's mission and about individual expectations. They follow the golden rule: They treat others as they would like to be treated. They show that they are competent and act in good intentions. They demonstrate that they're committed to the cause and are consistent in their actions.

CIOs also lead by their attitudes. IT team members look up to CIOs to see if they approach their roles positively and whether they tackle problems head-on.

Another trait of great leaders is that they are transparent. They don't hide information from members of their team, and they're clear about their intentions. They're also consistent in their messaging.

Great CIOs also exude confidence without being arrogant. They're driven not by ego but by the accomplishments of the team. Confident leaders admit their mistakes to senior executives and members of the team. A confident leader is constantly growing, absorbing new information and willing to accept change.

Great leaders also live by the same rules they set for others. Skirting the rules sends the wrong message to employees and what's expected of them versus senior management.

To help the enterprise succeed in an era of unprecedented change, CIOs have to draw on a new set of competencies around courage. This includes a willingness to go outside of their comfort zones and to lead with passion, conviction, and innovation.

Still, CIOs must also continue to demonstrate many of the leadership traits that enabled them to become the leaders that they are.

"I think today's CIO has to have a lot of qualities that 'old' CIOs use to have," says Rick Hopfer, CIO at Molina Healthcare in an HMG Strategy video. "You have to have good leadership skills, good relationship skills, communication skills. You have to be able to negotiate. You have to deal with conflict. But you also have to have a strategic vision. You have to be able to be adaptable. You have to know what's coming up and be able to communicate this in business terms. And you have to partner to deliver."

Redefining IT Teams in the Modern Enterprise

The radical pace of transformational change in business is leading to a number of dramatic shifts within the enterprise. Changes in customer expectations, including the digital touchpoints they use to research and purchase products, along with continuing disruption that's occurring across all industries, have forced companies to become more agile and responsive to changing customer behaviors and market requirements.

In order to respond effectively and provide the enterprise with the guidance, support, and services that are now needed, the IT organization must redefine the makeup and structure of its teams.

In the past, IT team members were recruited and onboarded based on their compatibility with the organization and the specific skills they brought to bear. But that's no longer sufficient. Today's IT teams must include a mix of internal and external resources that are able to bring the necessary competencies when and where they're needed. This includes the ability to bring in external resources from trusted business partners quickly to fulfill a specific skill set that's needed for a limited period of time.

CIOs also want people who can contribute different views and perspectives for tackling business and operational challenges. This is one of the benefits of recruiting and maintaining a diverse staff, each of whom brings a different background and set of experiences that shape their perspectives. Diversity can enhance creativity within the enterprise, casting a wider net of ideas for tackling business challenges and opportunities and for driving innovation.

Meanwhile, diversity can also aid CIOs with recruiting. In a recent Glassdoor survey, two-thirds of the people polled said that diversity was important to their evaluation of companies and job opportunities.

Diversity also factors into the viewpoints that different generations of employees bring to the table. For instance, millennials, or staffers who are 35 and younger, are widely regarded as *the connected generation*. They're creating, communicating, and collaborating differently than previous generations. The insights gained from millennial IT staffers can be used to help design workflows and user experiences that better accommodate the interests and needs of millennial workers.

As Mark Polansky, senior partner at Korn Ferry's Information Technology Officers Center of Expertise, points out in an HMG Strategy Transformational CIO blog post, the modern CIO assembles technology ecosystems that leverage the resources and expertise of both external partners as well as internal staffers.

"CIOs understand that a viable technology ecosystem is a complex web of vendors and service providers. It's very much like an M&A activity, and there are lots of moving parts and complicated relationships."

Business is moving and changing at the speed of light. CIOs need agile IT teams that are able to respond to shifting business requirements quickly with expertise and precision. Today's business initiatives can't be placed in an IT project pipeline and be delivered in weeks or months. To succeed in today's business climate, companies need the flexibility to move on emerging trends before the business has missed its opportunity to act.

Is It Time for a New Game Plan?

According to the Society for Information Management's IT Trends Study 2015, the average CIO is 51 years old, male (89 percent), and has been in his current position for about five years. Although none of this is terribly surprising, what is concerning is that the industry could lose up to one third to half of all current CIOs over the next 5 to 10 years, according to Leon Kappelman, the lead researcher for the report, in a recent interview with *Information Week*.

Although tenures for CIOs are longer than they've ever been, IT leaders don't last in their roles forever. Some CIOs move on to new IT leadership opportunities. Some take on new responsibilities in the C-suite, such as COO or CEO, while others move into consultative roles.

Regardless of the progression path, CIOs need to be thinking about laying the groundwork for next-generation

IT leaders. It's a carefully planned path that nurtures long-time rising stars within the IT organization, while drawing attractive talent from other organizations who can deliver needed skill sets as well as fresh perspectives to problem-solving. It requires conscientious planning on the part of the CIO to recognize, reward, and create new opportunities for valued employees while continuing to draw in external talent that can help the enterprise meet its goals going forward.

Certainly one of the biggest challenges CIOs face in nurturing next-generation IT leaders is retention. No one stays with a company for 20-plus years anymore. Gifted IT professionals are highly sought after. And they're more nomadic.

Compensation is just one piece of the puzzle. Competent IT professionals are also hungry for new opportunities, including stretch assignments that take them beyond their current skills or knowledge. These responsibilities can include overseeing projects that entail cutting-edge technologies or lines of business they've not worked with before. Stretch assignments provide learning opportunities for valued IT members while enabling the CIO to gauge how effectively an individual handles a particular project that falls outside of their normal comfort zone.

Of course, every CIO—and every employer—runs the risk of investing too much in a particular employee's career since they may very well take that knowledge elsewhere. Ultimately, IT organizations that don't provide employees with opportunities to take on new responsibilities and advance their careers will end up lacking the talent that's needed to help take the enterprise forward.

Meeting with IT staff regularly can help the CIO to understand their passions and to identify new creative opportunities that can help keep them energized. Great leaders don't place people in boxes—they free them from restrictions and give them room to run.

IT staff don't just want opportunities to grow their skills. They also want to share their observations and ideas and feel that their voices are being heard and are making a difference.

Great leaders create an environment where individual performers are nurtured and can spread their wings. They also lead by example by how they lead others, the relationships they build, and by how they're able to navigate change.

Ultimately, it's about cultivating an IT team that's able to deliver on the needs of the enterprise. As Ted Colbert, CIO at The Boeing Company, shares in an HMG Strategy video, "The modern CIO has got to look at their talent and make sure they have the right folks in place to deliver on the expectations of your business."

Partnering with Academia on STEM/STEAM/STREAM Development

Depending on the research source, the shortage of science, technology, engineering, and mathematics (STEM) professionals entering the US labor market ranges somewhere between acute and dire.

Economic projections are pointing to a need for 1 million more STEM professionals than the United States is currently producing at the current rate over the next decade, according to the President's Council of Advisors on Science and Technology. By the council's estimates, the United States would need to increase its annual production of undergraduates receiving STEM degrees by 34 percent over current rates to match the demand forecast for STEM professionals.

In 2015, computer and mathematical occupations employed roughly 4.7 million workers, according to the US Department of Labor, Bureau of Labor Statistics.

But it's not just a shortage of STEM professionals that IT executives are concerned with. STEAM is an alternative acronym that's gaining traction, which refers to science, technology, engineering, *arts,* and mathematics. By integrating STEM with the arts, academics and researchers believe that creative thinking can help young professionals with the problem-solving and deductive-reasoning skills that are needed to tackle the business and technology challenges in today's fast-paced, disruptive economy.

Taking this a step further is the importance of literacy skills. STREAM includes the reading/writing component. With the flood of data that's pouring in from a variety of sources (web, mobile, text, machine data, etc.), future IT/business professionals need the ability to be able to read, comprehend, analyze, and act quickly on large volumes of information.

While CIOs are looking for college grads with strong STEM/STEAM/STREAM skills, they also want aspiring tech professionals who have hands-on business experience and critical problem-solving skills. Increasingly, CIOs are expecting colleges and universities to provide undergrads with real-world training and learning environments so that they're ready to hit the ground running and not require a great deal of on-the-job tutoring. This could be accomplished, in part, by taking students through concrete business scenarios and projects they could work on, as well as internships and other immersive endeavors.

To help address these needs, CIOs can work directly with local colleges and universities that they actively recruit from to communicate the specific skill sets they're looking for. This can include evaluating course components and offering recommendations to strengthen their ability to apply IT knowledge to firsthand business challenges.

Moreover, CIOs can provide college students with real-world experience themselves by creating robust internship programs that enable students to obtain valuable work experience. The added benefit for CIOs is that they can also address short-term, entry-level project resource requirements while gauging whether a particular student is a good cultural fit for the organization long-term.

To help tackle the STEM/STEAM/STREAM labor shortages, CIOs can also become more actively involved with academia and associations to help engage young men and women well before they reach university age.

For its part, the Society for Information Management (SIM) and the Philadelphia Philanthropic Society for Information Management (PHISIM) both support a Teen Tech camp run by InspiriTec each summer. The camping program provides a fun and engaging environment that exposes teens to IT concepts and potential careers in technology.

Ultimately, it's about finding IT staffers who have a passion for what they do and aren't afraid to take risks. "We're looking for creative people who are not afraid of figuring out solutions," says Shobhana Ahluwalia, head of Information Technology at Uber in an HMG Strategy video. "We're just looking for people who go for it."

The Transformational Career Path

It wasn't that long ago that people were questioning whether the role of the CIO was still needed in enterprise companies. A lot has changed in recent years.

Digital skills and digital capital now constitute 22.5 percent of the global economy, according to Accenture. Companies have only just begun to unlock the potential for digital and need tech-savvy CIOs who understand how digital technologies can be applied to generate business value.

Moreover, as data have become the new oil, IT leaders are desperately needed to help companies determine which data sets offer the enterprise the most value and identify how raw data can be refined to help create new business opportunities.

These fundamental technology and market shifts are having a profound impact on the growing clout of the CIO. For instance, according to the Harvey Nash/KPMG CIO Survey 2016, which canvassed 3,352 CIOs and technology leaders across 82 countries, 34 percent of CIOs now report to the CEO, representing the largest portion in the survey's 18-year history. Meanwhile, 57 percent of CIOs surveyed in the study now serve as executive board members, compared to 51 percent last year.

As the CIO's responsibilities continue to evolve and be redefined, this is creating new career opportunities for CIOs. As business markets continue to be disrupted, the CIO has a new mandate as the CEO of technology and is being called on to provide bold approaches for leveraging technology to help companies drive business value and succeed in turbulent times. This manner of courageous leadership for challenging the status quo and forging strategic partnerships across industries is blazing a trail for fearless CIOs to ascend into higher leadership roles.

As companies push to digitize their businesses, there are growing opportunities for CIOs to transition into the COO role, according to Gartner, a US-based research and advisory firm specializing in IT issues. Savvy CIOs are able to draw on their experiences in building relationships and driving consensus across organizational functions to help digitization efforts succeed across the enterprise.

Meanwhile, as every enterprise has arguably become a technology company and as CIOs work closely with CEOs to architect the evolving business blueprint, this is creating new opportunities for CIOs to ascend to the CEO role. CIOs are well-positioned to leverage their comprehensive understanding of the business, their technology vision, and market acumen to help set a course for the twenty-first-century enterprise.

Ultimately, courageous leadership is about acting with passion, conviction, and commitment. "In terms of career ascent, you have to have relentless passion and commitment for what you're doing," says Jonathan Landon, Global IT—CTO at Kimberly-Clark Corporation in an HMG Strategy video.

Rethinking Career Development for Millennials

Digital disruption is forcing companies to support new ways to communicate with customers and CIOs to harness new IT skills and capabilities from their workforces.

For CIOs, millennials (recent college graduates to those in their mid-30s) represent a sizable portion of IT staffers who either already have these skills or require additional training to strengthen these skill sets.

Millennials are hungry to obtain new skills. According to a survey of 1,200 employed millennials conducted by Mindflash, 88 percent said they are willing to make personal sacrifices, including forgoing vacation time, in order to train themselves with the skills needed to compete in the workforce today.

As CIOs consider training options for millennials, it's important to recognize that many IT professionals from this generation have different training preferences and needs than baby boomers and gen X-ers.

A study conducted by Time Inc. found that millennials switch media types 27 times every nonworking hour. This demonstrates how millennials generally prefer to receive information and communications in small bits. Meanwhile, their penchant for multitasking reflects how they are nonlinear learners.

As high-level leaders in the enterprise, CIOs also need to identify and provide training for the skill sets that IT staffers need, including millennials. This includes pinpointing skill set gaps that can be addressed that can help the IT organization to perform at a higher level.

For millennials, this often includes soft skills such as communications, listening, and relationship building that is needed to be effective performers and to become successful leaders. According to a study of 592 business and learning professionals conducted by the Association for Talent Development (ATD), 56 percent of millennials are not equipped with the skills they need to be successful in the workforce.

Of course, not all employees are cut from the same cloth. Each staffer responds differently to various types of training.

Some Millennials, for instance, prefer shorter training sessions that include the use of digital tools that employees can make use of when it's convenient for them.

Many millennials also prefer informal training techniques. According to the ATD study, 53 percent of respondents find on-the-job training to be the most effective training and development strategy for millennials.

Millennials are also known to be workplace nomads, switching employers every two years on average compared to five years for gen X-ers and seven years for baby boomers, according to the ATD study. Although CIOs have to carefully consider the amount of training to invest in itinerant employees, job rotations are one way to help keep IT staffers engaged while helping to deepen the IT organization's bench strength.

CIOs need to train IT workers not only for the skills that are needed today but also to pave the foundation for the future IT organization. One study predicts that more than 3.6 million baby boomers will retire in 2017 and that one-fourth of all millennial workers will move into managerial posts. Identifying those IT staffers who exhibit leadership qualities and adequately preparing them for management roles has become more pressing than ever.

The dramatic increase in the use of digital tools and data is compelling CIOs to think differently about developing the skill sets needed to support the twenty-first enterprise.

"CIOs need to look at creating teams with the talent for building complex models and algorithms for machine learning, augmented reality and artificial intelligence," says Jim Fowler, CIO at GE, in an HMG Strategy Transformational CIO blog post. "Those are some of the areas where we'll be competing intensely over the next decade."

The Art of Self-Promotion

High-performing CIOs aren't always recognized for the breadth of their achievements. CEOs and other C-level executives who are stretched thin and have multiple direct reports and responsibilities can't keep track of everything that the CIO accomplishes.

This is just one of the reasons why CIOs need to promote themselves and their team's accomplishments. Making others aware of your accomplishments will create opportunities for promotions and salary increases. But that isn't the only rationale for self-promotion.

Raising awareness about the performance of the IT organization (and the CIO's leadership of the group) can help senior executives to more deeply appreciate the amount of value that the CIO and IT bring to the enterprise. Doing so is increasingly important in a landscape where some business leaders question the importance of having a CIO or otherwise make extensive use of external IT resources.

The IT organization can also benefit from a CIO's promotional efforts. The IT organization is only as strong as its people. IT staff who know that they have a CIO who is looking out for their best interests will reward the CIO with enthusiasm and loyalty.

CIOs who want to promote themselves without coming across as braggarts can focus on the work that's being done by the IT organization and the accomplishments that have been made. These triumphs can be communicated during presentations with the board, in IT steering committee meetings, and other forums that bring together senior executives. Presentations that tell a story, focus on outcomes, and call out the accomplishments of individual performers can highlight the value of the work that's been done and the people who are behind it.

Having current or former business colleagues to tout your achievements is another way to highlight a CIO's accomplishments with the added weight of having a respected peer endorse your feats. When a fellow C-level executive backs a CIO's contributions to the business on his or her LinkedIn profile, it's a great way of demonstrating a CIO's exploits to current (and future) employers.

Promoting the accomplishments of the CIO and the IT organization isn't completely self-serving. Sharing stories with the C-suite and line of business leaders about business projects that the IT organization has worked on, the challenges that were overcome, and the results that were achieved can be a way of creating awareness with senior management, while making real connections with people about the work that IT is doing with the business. It's also an effective way to build credibility with the C-suite and demonstrating your value to the business.

"The CIOs that are going to stand out going forward are not those who just sit at the table and wait for an idea to pop up that they can partner with the business to drive. But (those who) instigate and talk about the art of what's possible using technology to drive incremental value for the company and new markets for customers that the company may not be servicing today," says Steven Rullo, CIO at GE Capital–Treasury, in an HMG Strategy video.

No Executive Is Immune from Disruption

According to a 2015 study conducted by Harvey Nash and KPMG titled "CIO Survey 2015: Into an Age of Disruption," the majority of CIOs and other business leaders who were polled believe that not only will their organizations be impacted by digital disruption over the next 10 years but that the transformation that's taking place will provide their organizations with a competitive advantage.

The digital revolution is providing boundless opportunities for CIOs to shine as the spotlight on the execution of these efforts is being turned on them. Even though 34 percent of the respondents to the Harvey Nash/KPMG study say that the CMO is the primary executive leading digital business strategy compared to just 14 percent of CIOs, in many cases it's the CIO who is entrusted with implementing these efforts.

How successful CIOs are in enabling the enterprise to both circumvent and drive disruption will play a significant role in their future career paths. For instance, the rapid pace of change that's occurring in business is only going to accelerate moving forward. Those CIOs who are able to anticipate disruption from other forces and identify opportunities for driving disruptive innovation position themselves to step into future roles as CIOs, CEOs, and entrepreneurs.

Companies that have blazed new paths through disruptive innovation have wrested market share from other players, regardless of whether they're startups or industry incumbents. CIOs can lay the groundwork for capturing new customers and generating revenue growth by developing innovative business models that drive value both for customers as well as the enterprise.

Within any company, the CEO needs to understand each of the primary components that are essential to running the enterprise—operations, finance, sales, marketing, distribution, and so on. The CIO has become increasingly knowledgeable about each of these areas, as IT is embedded in all aspects of the business.

Andrew Rashbass is a shining example of this type of digitally fueled career rise. He was IT director at *The Economist* before becoming the CEO of the Economist.com. In an industry that has been completely revolutionized by digital disruption, Andrew has since become the CEO of *The Economist,* CEO of Thomson Reuters, and executive chairman of Euromoney.

Given the role and impact that digital technologies are playing in business today and going forward, future business leaders will need to be part-technologist and part-visionary. The new captains of industry will embrace how emerging technologies can be leveraged to craft new business models and capture market share.

Tim Stanley, president, CXO, and founder of Tekexecs/Innovatects/CXOco and former CIO at Caesars/Harrah's Entertainment advises in an HMG Strategy video, "The best thing you can do is to look at how you would disrupt yourself. And if you can focus on where are the threats, where are the opportunities, and what innovative things you would do if you were trying to displace you, then that gets you at least halfway to the game of figuring out what are the things you might want to do to drive innovation or stem that off against competitors."

Chapter 6

The "Eyes and Ears" of the Enterprise

As Trusted Senior Advisors, CIOs Serve Critical Roles

I had an excellent conversation with my good friend Patty Hatter, one of the most experienced and articulate technology executives I've known. Patty is senior vice president, Operations, and CIO at McAfee, which is now part of Intel Security.

We talked about the increasingly elevated role of the CIO as a trusted executive who captures critical insights and relays them meaningfully to the CEO and board senior management.

"Technology markets are continually evolving, and we often see the newest trends at their early stages," Patty says. "As CIOs, we tend to find out sooner what the other technology companies are offering, and that's incredibly valuable insight we can relay to senior management."

I genuinely appreciate how Patty frames the value of the CIO in terms of information flow. It makes complete sense that a company's IT leaders will be the first to learn of new technologies because they're constantly hearing pitches and presentations from technology vendors.

"We see what's out there, we find out what it can do, we learn about how it's priced, and we hear what other customers are saying about it," she says. "Lots of that information is highly valuable, and we try to learn from it."

In today's hyper-fast markets, it's difficult to stay current on the newest technology. Smart companies leverage the knowledge and expertise of their IT leaders to stay ahead of the competition. "You just can't assume that every good idea will come from within your own four walls. You've got to keep your eyes open and watch what's going on around you," she says.

Patty brings years of perspective to the evolving role of the CIO. Before joining McAfee, she was a vice president at Cisco, where she led the successful transformation of global processes and systems infrastructure. Overall, her team's efforts helped to enable and support Cisco's growth and scalability as net sales grew from $22 billion in 2004 to $40 billion in 2010 during her tenure.

At McAfee, Patty is responsible for driving cross-functional partnerships that accelerate delivery of strategic business priorities that impact bottom line profitability. She leads all facets of McAfee's ongoing transactional business and shared services, as well as IT, risk and compliance, and M&A integration.

From my perspective, Patty represents the new kind of CIO that I greatly admire: a trusted executive who plays a key role in guiding corporate strategy across turbulent and rapidly changing markets.

Nurturing Intellectual Curiosity to Assemble a World-Class Cybersecurity Team

Across industries, the shortage of qualified cybersecurity professionals in both the United States and globally has become acute. More than a quarter of enterprises report that the amount of time it takes to fill key cybersecurity and information security positions is at least six months, according to the Cybersecurity and the US State of the Union 2017 report from ISACA, which canvassed 633 cybersecurity and information security managers and practitioners.

Meanwhile, on average, 59 percent of enterprises receive at least five applicants for each open cybersecurity position, but most of these applicants are unqualified, according to the ISACA study.

Although demand for cyber professionals far outpaces supply, "We believe that we can teach techniques," says Mignona Cote, CISO PayFlex, senior director, Aetna Global Security. "What we can't teach is intellectual curiosity and so this is our primary focus in interviewing professionals at all levels."

One of the steps Aetna's Global Security team has taken to promote a culture of curiosity is by seeking cyber candidates who are committed to life-long learning. Mignona says, "We seek to learn what skill or competency the candidate wishes to pursue. We then build a plan for them to learn that specific skill in a given role, enabling them to invest their time with our resources to master the skill."

This often includes immersing each candidate in industry-leading techniques for security management. Every security employee allocates 10 percent of their professional time to attempt to learn something new through a research project or a working group.

"We call this 'play-time' and every individual is responsible for managing their play-time," she says.

The Global Security team also creates new roles for employees when they are ready to learn new skills and functional areas. Each employee has a professional development plan where they choose the skills they wish to invest in and Aetna Global Security provides vehicles for them to learn and apply these skills.

Quenching a Thirst for Knowledge

Mignona has found through her experiences that it's best to focus on candidates and employees that have a desire to learn new skills instead of the background talents of workers in determining whether certain people have an aptitude for cybersecurity. For instance, one of her most successful recruits was a marine biology major.

IT professionals also often show a strong aptitude for cyber skills, she says. "IT professionals often make great cybersecurity practitioners given their practical experience with IT in the enterprise," she says.

Her role as a security leader also lends itself as an entry point into security for new cyber team members. This includes her involvement with policy management, education and awareness, and security assessment. "It's a great way to show that non-IT person all the various aspects of security," she says.

For peers that are looking for creative ways to address the cybersecurity skills gap, Mignona offers the following recommendations:

- Know your people and build industry relationships.
- Spend time with your subject matter experts and give them the recognition and direction they need to continue growing in their roles.
- Keep your eyes and ears open for talent and create opportunities for them.

Looking beyond the Four Walls of IT and Seeing a World of Competitive Markets

Great CEOs want to hire CIOs who know technology. Great CEOs also want to hire CIOs who see beyond the walls of the IT department.

Can the needs of the CEO be resolved? Can CIOs simultaneously "keep the lights on" and help the company grow revenue? Can CIOs be technology gurus and strategy advisors at the same time?

I posed those questions recently to my friend Tom Peck. In his current role at AECOM, Tom wears several hats. In addition to serving as SVP and CIO, he is also the company's global head of procurement and travel. Tom really knows how to multitask effectively in a corporate environment.

"Of course, the CEO wants a CIO who can run the technology within the four walls of IT. That's how the CIO earns credibility," says Tom. "But the CEO also wants a CIO who can be a strategic advisor, someone who can recognize opportunities for using technology to grow the business."

For example, CIOs are usually the first C-suite executives to discover new technologies that could be turned into service offerings for their companies. CIOs can also provide invaluable technical insight for CEOs and boards of directors when their companies are considering mergers and acquisitions.

"Today, CEOs look for CIOs who can deliver technology *and* help the business monetize technology to create revenue. The modern CIO should be considered part of the sales team," says Tom. "The twenty-first-century CIO has broader responsibilities than CIOs of the past. Today's CIO is more involved in strategic initiatives. C-suite executives expect the CIO to get more deeply involved in the customer-facing side of the business. I definitely see that trend at our company."

From my perspective, Tom represents a great role model for the modern corporate CIO. Without a doubt, Tom is a CIO who sees above and beyond the traditional boundaries of IT. He fully understands that his job includes seeing the company in relation to markets it serves, and figuring out strategies enabling the company to become more competitive in those markets. CIOs like Tom are true members of the C-suite. They've earned their "seat at the table," and now they're demonstrating their unique value to the enterprise.

Great CIOs Know the Three Flavors of Disruption and How to Leverage Disruption to Create Value

I had a remarkably useful conversation with my good friend Snehal Antani recently regarding how to leverage disruption to create value for the modern enterprise. Snehal is the CTO at Splunk and the former CIO of North American Distribution at GE Capital. He has the kind of real-world, hands-on business experience that makes him a truly valuable IT executive.

I asked him for his advice on dealing successfully with the boards of major corporations, and he began by describing the five areas of deep knowledge and expertise that board members typically bring to the table: risk management, disruption, transformation, talent management, and capital allocation.

Since each member of the board is a black-belt master in one or more of those areas, you can anticipate the kinds of questions they're likely to ask and prepare in advance.

Of those five areas, the "hottest" right now is probably disruption. So it's critical for CIOs to understand that not all forms of disruption are equal, relevant, or desirable.

"As a technology expert, the CIO is uniquely qualified to serve as the voice of disruption," Snehal says. "But disruption comes in three different flavors, and you need to know which flavor is right for your environment."

The first flavor of disruption is exemplified by the kinds of "moon shot" projects undertaken by visionary entrepreneurs like Elon Musk. The second flavor of disruption is targeting monopolies and disrupting their markets. For Global 500 firms, however, those "flavors" are not usually on the menu. "Moon shots" are risky and can take decades to generate profits, which makes them unappealing for most large companies. If you're a Global 500 firm, you probably are one of the largest players in your markets, so disrupting those markets doesn't make sense.

The third flavor of disruption is integrating and optimizing a broken or fragmented value chain. That flavor of disruption is most likely the best choice for large companies with extended supply chains and global markets. Companies such as Apple and Walmart have proven beyond a shadow of doubt that value chain integration can be an incredibly effective form of disruption—and it can be extremely profitable, too!

Apple, for example, integrated a highly fragmented music industry by offering a single platform for consumers. Walmart revolutionized retailing by applying big data analytics to its entire supply chain, from sourcing raw materials and products to stocking shelves and accurately anticipating consumer demand.

The best and brightest CIOs understand how to deliver meaningful disruption to the modern enterprise. They understand that disruption must have a purpose, and they develop the executive-level communication skills necessary to explain the business value of disruption to the board.

Twenty-First-Century CIOs Need Split Personalities to Balance Continuity and Growth

In ancient Roman culture, the mythical deity Janus was depicted with two faces: one facing the past, the other facing the future.

Janus had the rare ability to look in two directions at once, and understand what was happening on both sides.

From my perspective, Janus would be an excellent role model for the twenty-first-century CIO.

In the modern enterprise, you can't afford to focus solely on the future or on the past. You need the ability to shift your focus rapidly and comprehend the big picture.

The idea of looking in two directions at the same time arose during a wonderful conversation with Kevin Sealy, senior client partner and EMEA CIO Practice Head at Korn Ferry. Kevin suggests that the modern CIO needs a *split personality*—on the one hand, the energetic and innovative persona driving investment in new digital systems, and on the other the individual who ensures five nines availability and who is still trying to rationalize hundreds of legacy applications and refresh an underinvested infrastructure.

"You're seeing a demand for both styles in CIO appointments," says Kevin. "On the one hand, companies want a CIO with an innovative technology agenda. But on the other hand, companies want someone who can rationalize the legacy and dramatically overhaul the cost base."

While it might be tempting to forget the legacy systems, the hard truth is that many of those older systems are not just crucial for day-to-day operation, but are themselves a critical foundation for newer digital front-ends, if managed and standardized effectively.

That's why many companies are looking for CIOs who can work with both legacy and newer systems. "As the CIO, you're still required to pull together technology, business strategy, and data. That hasn't changed," says Kevin. But most companies also want their CIO to be a proactive agent of change, a top-notch senior executive who can guide the enterprise smoothly into the future.

"Today, the C-suite is much more aware of technology. They know about tech because they use it in their everyday lives. And they want to know how they can use the tech around them to drive new revenues, open new channels for sales and compete successfully in new markets," says Kevin.

Kevin has definitely hit the proverbial nail on the head. The modern CIO needs that unique ability to look in two directions at the same time, and strike a good balance between continuity and disruption. Continuity keeps the business going, but disruption makes it grow.

Fully Engaged IT Leaders Drive Business Growth and Value in Highly Competitive Markets

I speak often with my good friend and colleague, Ramón Baez, the former senior vice president and chief information officer (CIO) at HP. In his role at HP, Ramón was responsible for the global IT strategy and all of the company's IT assets, which include worldwide application development, the company's private cloud, IT security, data management, technology infrastructure, and telecommunication networks.

Ramón embodies the "CIO as CEO" concept, and he sets the standard for today's generation of IT leaders who are fully engaged and actively helping their companies grow top line revenue in competitive markets.

"In the world of IT, we're moving faster than ever before and we're becoming much more service oriented," Ramón says. "We are driving innovation and leading change. When you look under the hood, you see that IT has become a critical enabler of growth. We're a key part of the business, and we enjoy being an engine for innovation."

Ramón's approach contrasts sharply with IT strategies of the past. In the early days of IT, it was normal for CIOs to resist change and shun risk. But those days are gone. IT is no longer the "office of no." Instead, it's the place where business leaders go to find support for projects that will create competitive advantages for their products and services.

"Companies that can leverage technology will create new opportunities for themselves and disrupt their competitors," Ramón says. "In modern markets, it's much better to be the *disruptor* than the *disruptee*."

You also have to be fast to compete in today's rapidly changing markets. In the old days, it was common for IT to wait six months or a year to update an application. Some applications were updated biennially, which seems hard to imagine today.

Nowadays, many applications are updated continuously. That's the new normal—continuous change and improvement. "We're constantly creating new applications that help us compete against other companies," Ramón says.

Speaking with Ramón always raises my levels of optimism about our industry. I am totally confident that leaders like Ramón are guiding us to a future state in which IT is more integrated, engaged, and involved in generating real business results than ever before. I've been writing for years about IT's potential for driving business growth, and I'm absolutely delighted that many of my predictions have proven accurate.

Intel Takes Another Look at Gordon Moore's Famous Law

Despite its edgy title, *Only the Paranoid Survive* is really more about intelligence than instinct. I never had the opportunity to meet Andy Grove, Intel's legendary chairman of the board, but there's no question in my mind that he was totally brilliant and a great business leader.

My guess is that he picked the title because he knew it would help propel his book quickly to the top of the bestseller lists. From my perspective, it was a savvy business decision.

I firmly believe that Andy, who passed in 2016, would be among the first to agree that you always need to test your assumptions and rethink your business plans. I was fascinated when Brian Krzanich, Intel's current chief executive, said the company is no longer dominated by Moore's law—the famous concept invented by Gordon Moore, Andy's long-time colleague.

"Silicon Valley treats Moore's law as if it is immutable, and with even more reverence than it does paranoia. But it was not a scientific law; it was always an observation about the behavior of a market for computers and software, which paid off at a rate to justify increasing investment in making chips," Quentin Hardy wrote in a recent column.

As Hardy noted, Intel is focusing now on wireless and networking, segments of the tech market that seem indifferent to Moore's law. Intel is also renewing its push to create incredibly powerful new memory chips with the potential to radically transform computing.

From my perspective, that sounds more like great innovation than wild paranoia. Intel has always been a smart company, and I think that Andy's intellectual drive and curiosity will always be a major part of Intel's corporate culture.

Interacting Successfully with Senior Management

Deanna Wise is EVP and CIO at Dignity Health, formerly Catholic Healthcare West. I spoke with Deanna recently, and our conversation reminded me of the critical importance of blending technology expertise with business acumen and excellent diplomatic skills.

We all know that healthcare is a challenging field, and it offers great lessons for CIOs in all industries. Deanna told me a great story about how she handled a potentially difficult situation in a previous job by finding the right balance of people, process and technology.

In her anecdote, Deanna relates how the company had recently changed its name. As a result, the CEO wanted every report in the company changed to reflect the new name within one week's time. For someone unfamiliar with the processes of IT, that might seem like a perfectly reasonable request. But those of us with IT experience know first-hand that fulfilling such a request can easily require hundreds of hours of labor.

From my perspective, Deanna devised a solution that was both elegant and efficient. She explained to the CEO that immediately changing every report would be an expensive and time-consuming project. She offered a reasonable alternative: Reports that were circulated outside the company would be changed immediately. Internal reports and other noncritical documents would be changed on a rolling basis, as needed.

"He was adamant that he needed the reports changed," Deanna recalls. "But when I showed him the actual costs of hiring an outside consultant to change all of the reports within one week, it was easy for him to make the right decision. When you present the information and you give people a choice, it's usually easier to get the outcome you want."

I genuinely admire how Deanna handled that situation, using facts and numbers to make her case. I think it's a good lesson for all of us in the IT field. We often tend to assume that everyone in the C-suite understands technical issues and costs. The simple truth is that most executives don't fully understand the technology or the costs involved with IT projects, so you really need to bring your best diplomatic skills to the table.

Deanna's story also demonstrates why it's important for the CIO to resist a call for sweeping change when a series of incremental changes will produce a better outcome for the enterprise.

I really appreciate how Deanna finds the right balance in her interactions with the C-suite and executive board. Everyone benefits from mutual respect, and it's always best to strive for win-win outcomes.

"We Have to Be Very Comfortable with Being Uncomfortable"

When I think about courageous leadership on a truly global scale, an example and role model who comes to mind is Kevin Haskew, senior vice president and chief information officer at ON Semiconductor, a Fortune 1000 company offering a broad and comprehensive portfolio of innovative technology products.

I sat down recently for an in-depth conversation with Kevin, who spoke with me about his approach to effective leadership in turbulent times. In our conversation,

Kevin emphasized the need for CIOs and other technology executives to hone their skills in communicating with the C-suite and boards of directors.

"You've got to be seen as credible," says Kevin. "You can't just talk IT. You need to communicate from a business perspective. Ideally, they should see you as a businessperson. It helps if you've had prior P&L responsibility."

Kevin says it's important to remember that many senior-level executives are still wary of technology and don't really understand its role in the business. "A lot of executives are still nervous with technology," he says. "Sometimes, they feel threatened or lost, and it makes the relationships a little bit uncomfortable."

Even when you need to move forward quickly into the future, "you still have to make sure that the day-to-day operations are covered and that your systems are running smoothly, efficiently, safely and securely," says Kevin. "If they're not, then you've got no credibility."

The good news is that when you've got good communication skills and you're perceived as a credible leader, "then people start listening to you. You can start to have those conversations about changes in the industry, new technologies, new directions, and new strategies," says Kevin.

Kevin also says that it's important to step out of your comfort zone and keep up with unexpected shifts in markets.

Expectations are always changing, and tech leaders need to stay aware of how consumers think and how they behave. In chaotic twenty-first-century markets, it's absolutely critical to keep the customer experience "top of mind" at all times.

"As executive leaders today, we have to be very comfortable with being uncomfortable," says Kevin. "That's the way it is."

I genuinely appreciate and value Kevin's excellent advice. He truly is an outstanding role model and a trusted source of guidance in difficult times.

Great CIOs Create Highly Effective Teams to Drive Value for the Modern Enterprise

Great CIOs create highly effective teams of exceptional vendors and service providers to drive real value for the modern enterprise.

I had an excellent conversation with my good friend Mark Polansky recently, and I want to share some of the highlights with you. Mark is a senior partner in *Korn/Ferry International's* Information Technology Officers Center of Expertise, a practice he has led for more than 12 years.

My conversation with Mark focused mainly on one of our favorite topics: the continuously evolving role of the CIO in the modern enterprise. One area that is often overlooked is the CIO's increasing responsibility for assembling teams of external partners to accomplish mission-critical tasks.

In the past, most of those critical tasks, in addition to the technologies required to support them, would have been handled in-house by the IT department.

As we all know, however, modern markets move at extreme velocities. Today's CIO cannot simply tell the business to wait 18 months while IT builds a homegrown solution. By the time a homegrown solution is built and deployed, the business will have moved on to a new set of challenges.

Smart CIOs operate like M&A (mergers and acquisitions) specialists. They have to know which vendors provide the right services and which vendors they can trust. They also need the leadership skills for managing teams of vendors over extended periods of time, under highly fluid and often unpredictable conditions.

"The modern CIO puts together technology ecosystems," Mark says. "CIOs create and manage partnerships between hardware, software, and service providers. They understand that a viable technology ecosystem is a complex web of vendors and service providers. It's very much like an M&A activity, and there are lots of moving parts and complicated relationships."

I really like how Mark describes the twenty-first-century technology ecosystem as a vast and complex web of companies. Managing those kinds of ecosystems requires exceptional leadership skills and a deep understanding of how different kinds of technology interact in real-world situations.

"Today's CIOs cannot sit on their laurels and they cannot deliver services the old-fashioned way," Mark says. "The mantra now is that if you can't digitize and disrupt your own business, you're leaving the door open for a competitor who is faster and more agile than you are. Then you will be playing catch-up, and that's usually a losing game."

The image of the CIO as a team builder fits perfectly with my concept of the CIO as the new "CEO of technology." When you think about it, great CEOs are almost always great team builders and team leaders. They understand that no single part of the enterprise can get the job done all by itself—success requires all hands on deck, working together smoothly and effectively, under difficult circumstances.

Chapter 7

Building a Robust Network of Partners

Leveraging the Partner Network for Digital Transformation Success

One of the biggest challenges associated with digital transformation is determining the right steps the company should take on its digital journey and which actions to prioritize.

It's not just neophytes that are struggling to plot their digital strategies. According to the 2016 State of Digital Transformation report conducted by Altimeter Group, just 29 percent of companies have a multiyear roadmap to guide their digital transformation evolution.

Many decision-makers begin the digital journey by identifying the company's primary challenges and customer pain points and exploring how digital initiatives can be applied to help address those. Others focus on the organization's short- and long-term business objectives (e.g., increase incremental revenue, boost customer loyalty) and mapping digital efforts to those goals.

Savvy CIOs have approached digital transformation as an opportunity to explore *the art of the possible*—identifying services, or business processes that digital technologies can empower that weren't before possible or feasible.

Another pragmatic approach to digital transformation is tapping into insights from vendor partners. Enterprise vendors have vast experience working with companies on a variety of digital projects across multiple industries.

Close-knit partners that have a view into your company's operations and culture can offer recommendations on best practices and approaches that have worked for other clients for comparable applications. Strategic partners can also share tactics that other clients have applied to tackle some of the thornier challenges of digital transformation, such as steps taken to foster the cultural adoption within the employee base necessary to achieve success.

For instance, a CIO for a bank may discover through his or her relationship with a digital technology partner how a retailer is gathering, analyzing, and acting on digital customer behavior in order to generate the next best action for high-value customers.

Meanwhile, a technology partner may also share how clients in other industries were able to cut the number of steps required in business processes in their shift to a digital model. Or how digitization has been able to streamline the number of documents needed for approvals to be made.

As digitally experienced CIOs recognize, digital transformation can generate a number of ancillary benefits. For instance, replacing paper and manual processes with software and digital tools can enable companies to collect valuable data that can be more easily analyzed to detect changes in business conditions or causes of risk.

While networking with peers can also help to yield these types of insights, having a robust network of trusted technology partners can further aid in identifying potential use cases and cost-savings opportunities achieved through digital transformation.

"That's what transformation really is—it's about people," says Scott Schwarzhoff, VP, Product Marketing, Okta in an HMG Strategy video. "As we move from a world of on-premise software into the cloud and into things like mobility technologies and challenges such as security, from a CIO's perspective, it becomes about partnership relationships with their employee-facing organization, their partner-facing organization, and their customer-facing organization."

Playing a Central Role in Enterprise Transformation

Digital disruption and business model innovation are forcing enterprise companies across industries to transform their businesses to improve competitive standing. Thanks to emerging competitive threats, companies can no longer rely on traditional business practices to retain customers and market share. A study by INNOSIGHT predicts that 75 percent of the S&P 500 will be replaced by 2027.

In short, companies need to find new ways to deliver value to customers to drive sustainable growth.

In the aviation industry, competition is fierce. As aircraft makers look for new ways to bring innovation to design while improving fuel conservation aimed at helping airlines to reduce their operating costs, manufacturers are increasingly relying on data and digitization to help drive new business breakthroughs.

One of the leaders behind the digital data transformation in the aviation industry is Ted Colbert, CIO and SVP, Information and Analytics at The Boeing Company. Colbert is an HMG Strategy 2016 Transformational CIO Leadership Award winner, having been recognized along with other top-performing CIOs and technology executives for excelling in leadership, innovation, transformation, industry give-back, management of extreme growth, and legacy leadership.

Colbert is leading Boeing's business transformation efforts by utilizing the flow of digital data throughout the lifecycle of the company's products. The data-driven approach is enabling the aviation company to improve its products and glean insights from design and manufacturing to help its customers run their businesses more effectively, as described by *CIO.com*.

Where there were roughly 100 sensors installed in the nose cone of a Boeing aircraft built in the 1960s, there are now thousands of sensors spread across each plane that gather vital data that can be used in design and manufacturing and shared with customers. For instance, the data can be used to predict part failure, determine when maintenance needs to be scheduled on an aircraft, and optimize crew scheduling.

Another long-established company that's succeeding with its digital transformation efforts is GE. The multinational conglomerate is investing heavily in Internet of Things (IoT) technologies to help create new products, improve services, and generate efficiency gains in the industries where GE is active—such as energy, healthcare, and transportation.

In 2011, the company created a software center in San Ramon, California, where one of the top projects is to build an operating system for factories and industrial equipment, according to *The New York Times*. GE considers the project central to the company's goal of becoming a "top 10 software company" by 2020.

To help GE execute on this vision, CIO Jim Fowler is focused on putting together teams with the types of cutting-edge skill sets needed for the company to succeed in the new economy.

"CIOs need to look at creating teams with the talent for building complex models and algorithms for machine learning, augmented reality, and artificial intelligence," Fowler said in an HMG Strategy Transformational CIO blog post. "Those are some of the areas where we'll be competing intensely over the next decade."

In today's highly competitive business landscape, CIOs also need to forge strong connections with business leaders and technology partners to brainstorm and identify new opportunities for driving digital transformation. This includes understanding the key business objectives that organizational leaders are striving to accomplish and to distinguish ways in which digitization can be used to create business value and move the enterprise forward.

Meanwhile, the CIO can also leverage his or her relationships with key technology partners to discover how clients in other industries are harnessing digitization along with the lessons that can potentially be applied to the company's business strategies.

Don't Underestimate the Difficulties of Digital Transformation

If digital transformation isn't at the core of your organization's business strategy, it should be. As a growing volume of consumers and other businesses increase their use of digital channels to conduct research on products and services and to complete transactions, digital business opportunities are skyrocketing for enterprises across vertical industries.

According to an Accenture study, 58 percent of businesses now look to digital to help them sell profitably. Still, many enterprises have miles to go in their digital maturity journeys. The same study found that just 26 percent of companies are "completely ready" to execute on digital business strategies.

Meanwhile, a separate study of 422 business executives conducted by YouGov on behalf of Appian found that while business leaders are aware of the need for digital transformation, only 14 percent have "fully migrated" to all intended areas of their digital transformation plans.

Part of the challenge facing enterprises is that many nontechnical business leaders don't know where to begin in launching digital business strategies due to their lack of familiarity with digital tools and their capabilities.

The CIO is well-positioned to help the C-suite paint a vision for digital strategies, starting with a broad brush and identifying key goals (e.g., driving greater online revenue with customers). These tactics should include discussions about what the company's short- and long-term digital goals should be and how they need to be aligned with the organization's short- and long-term business objectives.

Thanks to their knowledge with digital tools and their capabilities, the CIO is also perfectly situated to help the enterprise to develop a digital roadmap, including the tools, people/skills, and processes that are needed to execute effectively on digital business strategies.

While selecting the right tools to meet organizational objectives is essential, it's equally important to develop and recruit people with the right skill sets to enable the organization to succeed in its digital efforts. By leveraging their view across the various functions and business units across the enterprise and how they interoperate with one another, the CIO can work with HR and business/functional leaders to identify the types of skill sets that are needed to help different parts of the enterprise execute on its goals.

For instance, if a company's goal is to grow its customer base and revenues, the CIO can work with organizational leaders to identify the types of digital marketing skills that are needed to help attract targeted customer segments.

The CIO also plays a critical role in forging partnerships with the right third-party providers that can add value to the organization's digital business strategy. Enterprises can't execute on digital business strategies by themselves. The CIO needs to act with courageous leadership to identify those partners that share a vision for the company's digital future while meshing with its digital roadmap.

Once the enterprise has installed the building blocks for digital business, companies also must apply the right metrics and measurements to gauge how effectively the organization is executing on digital business strategies. Since customers often use multiple touchpoints to conduct research and make decisions for purchasing a product or service, measurements should include the impact of customer engagement across multiple channels.

The CIO should also ensure that senior executives are receiving regular updates on the progress of digital business initiatives. This can include the use of dashboards and other tools that can be used to provide senior executives with timely reports that contain key performance indicators.

For CIOs, their roles in driving digital business strategies is about identifying and executing on ways that they can deliver value to the enterprise. Conventional approaches are easy, but they aren't enough for today's highly demanding business environments.

Driving Business Transformation via Process Revolution

Three core elements are required for driving transformational change: people, processes, and technology. There are many writings on the role of the CIO in executing on leading and managing people—including building relationships and fostering the cultural changes that are needed for business transformation to occur.

There are also technological components involved in achieving transformation, including the distinctive ways that CIOs are leading digital disruption. These elements include the ability to communicate a clear vision for digital transformation to senior management and how savvy CIOs are highly engaged in helping lines of businesses develop new products and services.

While each of these competencies is essential for success, processes are often the mysterious ingredients that harmoniously bind the other elements together.

The CIO's role in the enterprise isn't solely to identify and ensure the successful delivery of technology and technology services to different constituents. The CIO is also counted on to grasp how various organizational functions and business units interoperate and rely on one another to successfully deliver products and services to end customers.

A simple example of this is a manufacturer that needs to have its order management, supply chain, procurement, and financial reporting processes coordinated together successfully in order for its supply chain and distribution systems to operate effectively.

The manufacturing supply chain analogy is admittedly old school. In order for digital transformation to occur successfully, business processes often need to be radically overhauled. For instance, in recent years, the balance of power has shifted to customers. Customers are driving the conversation with companies—they've become more vocal about their interests and preferences through the power of social media. As a result, this has fundamentally changed how organizations need to engage with them.

As a result, customer-facing processes have to be revolutionized to cede control to customers. Businesses need to develop new processes that support customers' use of digital devices, social media, self-service technologies, and other touchpoints that customers now rule.

Meanwhile, as the pace of business has gone into overdrive, faster, more agile processes are needed to accommodate the needs of consumers and business partners. Because of their unique view of the enterprise, CIOs are well-positioned to offer advice on processes that can be streamlined or even eliminated.

One of the ways the CIO can deliver value is by identifying tools and techniques that can be used to evaluate and act on existing or new processes that are needed. For instance, customer journey mapping tools can be used by marketers and other customer experience leaders to plot out the steps that target customers take in their path to purchase. By charting these steps, business leaders can identify processes and other points of friction that can be improved or removed.

"Disruption is pivotal in the role of the CIO," says Robin Brown, VP and CIO at Mortenson in an HMG Strategy video. "CIOs really need to be the bridge between business partners and the IT team. Understanding where your gaps are—that's where you need to create a lot of disruption."

In some cases, business and operational processes represent the gaps where CIOs can apply disruption. Business transformation relies on the three-legged stool of people, processes, and technology. If one leg of the stool isn't correctly aligned with the others, the entire stool will be off-balance and won't function properly. The CIO plays an essential role in overseeing the design and execution of business transformation.

Chapter 8

Key Takeaways

We learned many valuable lessons as we conducted our research for this book, the fifth in a series of leadership books conceived and created by the team at HMG Strategy LLC. Here are some of the main points and key takeaways from our comprehensive and truly unique method of in-depth peer-to-peer research:

- Just 26 percent of companies are "completely ready" to executive on digital business strategies, according to Accenture.

- Thanks to their familiarity with digital tools and their capabilities, the CIO is well-positioned to help the enterprise develop a digital roadmap, including the tools, people/skills, and processes that are needed to execute effectively on digital business strategies.

- The CIO also needs to act with courageous leadership to identify technology partners that share a vision for the company's digital future while meshing with its digital roadmap.

- Although people and technology are critical components for achieving business transformation, essential for success, processes are often the mysterious ingredients that harmoniously bind the other elements together.

- The CIO is counted on to distinguish how different functions and business units interoperate and rely on one another to successfully deliver products and services to end customers.

- Another way the CIO can deliver value is by identifying tools and techniques that can be used to evaluate and act on existing or new processes that are needed.

- The CIO's technology/business expertise is sorely needed by the board of directors. According to a 2016 study conducted by Accenture, just 10 percent of board members on corporate boards have professional technology experience.

- The first thing that boards look for from CIOs is the ability to speak clearly about technology issues in business terms.

- One way that CIOs can keep board members up to speed on the state of cybersecurity is by sharing examples of cyberattacks that have been launched against other companies, along with the impact that these attacks have had on operational uptime, margins, customer loyalty, and so on.

- Despite enormous opportunities for driving business value and operational improvement through digital transformation, there's still confusion regarding who should own and drive digital strategies within the enterprise. According to a Forrester/Accenture study, digital ownership is currently divided between the CEO (38 percent), the CIO (33 percent), and other senior leaders such as the chief digital officer and the chief marketing officer (29 percent).

- One of the ways that the CIO can demonstrate leadership here is by sharing his or her insights regarding areas of the business or specific operational activities that could benefit from digital deployment based on their comprehensive view of the enterprise and how all of the various business lines and organizational functions are connected and interoperate.

- The CIO can also act as the chief partnership officer to help the enterprise identify those business/technology partners that offer the best fit for the company's digital vision.

- Because many companies are struggling to plot the course of their digital transformation journeys, feedback obtained from trusted technology partners that have worked with clients across industries can provide valuable insights.

- Technology partners can share potential use cases as well as tactics for tackling top challenges to digital transformation, such as approaches for obtaining widespread cultural adoption.

- Trusted partners can also share examples of how clients in various industries have applied digital initiatives to streamline business processes, cut costs, or generate new revenue streams.

- CIOs aren't leaders because of their title but because of the actions that they take. This includes how CIOs handle decision-making, whether they act decisively, listen to and act on input from managers and staff, and how they communicate the actions that have been taken.

- Great leaders are highly trusted, and they gain this trust through their competence, through clear communications, through their consistency, and by how they treat others.

- Exceptional leaders are driven not by ego but by the accomplishments of the team, and they readily admit their mistakes.

- The rapid pace of transformational change is forcing CIOs to redesign and structure of IT teams.

- CIOs that create diverse IT staffs can benefit from broader perspectives for tackling business challenges and driving innovation.

- Savvy CIOs are creating ecosystems of external and internal expertise to fulfill their skill set requirements.

- There are multiple benefits that can be generated from the deployment of test labs and sandboxes. These include opportunities for IT team members to gain hands-on experience working on next-generation innovation efforts that have the potential to add meaningful value to the business.

- Test labs offer opportunities for the IT organization to seek out business problems and opportunities for disrupting markets and catering to unmet customer needs.

- True Innovation doesn't always emerge from a radical change or a disruptive action.

- One of the ways a CIO can help drive incremental innovation is by lending their project management expertise to ensure that project teams are staffed by people with the right skills and stick to project timelines.

- Thanks to their understanding of existing and emerging technologies, the CIO can also help identify opportunities for applying technology to improve internal or external business processes.

- Companies are reaching outside of their traditional supply chains to collect additional ideas for innovation, including customers and other outside contributors.

- CIOs can promote the benefits of open innovation by sharing examples of other companies that have jumped ahead of the market and gained competitive advantage.

- CIOs can also work with a respected business leader who can champion open innovation and influence the C-suite on its merits.

- As a chief process coordinator within the enterprise, the CIO is responsible for recognizing fresh ways for managing innovation efforts.

- The CIO can prompt innovation teams to explore other potential uses for a product that could lead to greater revenue potential.

- Instead of simply considering a new product, examine opportunities for bundling in services that can create an entirely new business model.

- CIOs must pay close attention to how the business defines innovation and then align IT to strategy to match those definitions.

- The CIO must also understand where the business is headed, the markets it is trying to penetrate and capture, and then work with the business to identify solutions to achieve those goals.

- Thanks to their unique view across the enterprise, the CIO can serve as a disruptive innovator.

- According to recent research from the Society for Information Management (SIM), one-third to half of all CIOs could be retiring over the next 5 to 10 years.

- Forward-looking CIOs need to recognize, reward, and create new opportunities for valued employees while continuing to draw in external talent that can help the enterprise meet its objectives.

- Stretch assignments provide learning opportunities for valued IT employees while enabling the CIO to gauge how effectively an individual handles a particular project that falls outside of his or her normal comfort zone.

- Industry projections point to a need for 1 million more STEM professionals than the United States is currently producing at the current rate over the next decade.

- Demand is also growing for STEAM (science, technology, engineering, arts, and mathematics) and STREAM (with "R" representing reading/writing) professionals who can draw on their artistic acumen to attack problem-solving creatively and to gather, analyze, and act on large volumes of data effectively.

- To help address these needs, CIOs can work directly with local colleges and universities to identify and recommend approaches for incorporating real-world experiences and skills into their curricula.

- Foundational technology and market shifts are creating new leadership opportunities for the CIO.

- As companies increasingly digitize their businesses, the operational and strategic requirements for creating digital business models are creating new opportunities for CIOs to transition into the chief operating officer (COO) role.

- CIOs are well-positioned to leverage their comprehensive understanding of the business, their technology vision, and market acumen to set a course for the twenty-first-century enterprise.

- Digital disruption is forcing CIOs to harness new skills and capabilities from their IT staffs.

- As CIOs consider training options for millennials, they must recognize that many IT professionals from this generation have different training preferences and needs than baby boomers and gen-Xers.

- CIOs also must identify the skill set gaps that need to be addressed to help the IT organization to perform at a higher level.

- CEOs and other C-level executives who are stretched thin and have multiple direct reports and responsibilities can't keep track of everything that the CIO accomplishes.

- Raising awareness about the performance of the IT organization (and the CIO's leadership of the group) can help senior executives to more deeply appreciate the amount of value that the CIO and IT bring to the enterprise.

- Having current or former business colleagues tout your achievements through LinkedIn recommendations and other forums is another way to highlight a CIO's accomplishments with the added weight of a respected peer offering their perspective.

- Digital disruption is providing incredible opportunities for CIOs to shine as they take on responsibility for execution.

- CIOs can lay the groundwork for capturing new customers and generating revenue growth by paving innovative business models that drive value both for customers as well as the enterprise.

- Evaluate how your own organization is vulnerable to disruption to identify both the threats and opportunities for guarding against and driving disruption.

- Companies can no longer rely on traditional business practices to retain customers and market share.

- CIOs also need to forge strong connections with business leaders and technology partners to brainstorm and identify new opportunities for driving digital transformation. This includes understanding the key business objectives that organizational leaders are striving to accomplish and to distinguish ways in which digitization can be used to create business value and move the enterprise forward.

RECOMMENDED READING

Adner, Ron. *The Wide Lens: What Successful Innovators See That Others Miss*. New York: Portfolio/Penguin, 2012, 2013.

Amoroso, Edward G., and Matthew E. Amoroso. *From Gates to Apps*. Summit, New Jersey: Silicon Press, 2013.

Collins, Jim. *Good to Great: Why Some Companies Make the Leap ... and Others Don't*. New York: HarperBusiness, 2001.

Covey, Stephen R. *The 7 Habits of Highly Effective People: Powerful Lessons in Personal Change*. New York: Free Press, 1989, 2004.

Christensen, Clayton M. *The Innovator's Dilemma: The Revolutionary Book That Will Change the Way You Do Business*. New York: HarperBusiness, 1997, 2000, 2003, 2011.

Domingos, Pedro. *The Master Algorithm: How the Quest for the Ultimate Learning Machine Will Remake Our World*. New York: Basic Books, 2015.

Greenleaf, Robert K. *Servant Leadership: A Journey into the Nature of Legitimate Power & Greatness*. New York: Paulist Press, 2002.

Katzenbach, Jon R., and Douglas K. Smith. *The Wisdom of Teams: Creating the High-Performance Organization*. Boston: Harvard Business School Press, 1993.

Kotter, John P. *John P. Kotter on What Leaders Really Do*. Boston: Harvard Business School Press, 1999.

Krishnan, M.S., and C.K. Prahalad. *The New Age of Innovation: Driving Cocreated Value Through Global Networks*. New York: McGraw Hill, 2008.

Lencioni, Patrick. *The Five Dysfunctions of a Team: A Leadership Fable*. San Francisco: Jossey-Bass, 2002.

Maxwell, John C. *The 21 Irrefutable Laws of Leadership: Follow Them and People Will Follow You*. Nashville: Thomas Nelson, 1998.

Newman, Sam. *Building Microservices: Designing Fine-Grained Systems*. Sebastopal, California: O'Reilly Media, 2015.

177

Rossman, John. *The Amazon Way: 14 Leadership Principles Behind the World's Most Disruptive Company*. Seattle: Clyde Hill Publishing, 2016.

Schein, Edgar, H. *The Corporate Culture Survival Guide*. San Francisco: Jossey-Bass, 1999.

___. *Helping: How to Offer, Give, and Receive Help*. San Francisco: Berrett-Koehler Publishers, 2009.

___. *Organizational Culture and Leadership (Fourth Edition)*. San Francisco: Jossey-Bass, 2010.

Watkins, Michael. *The First 90 Days: Critical Strategies for New Leaders at All Levels*. Boston: Harvard Business School Press, 2003.

Zweifel, Thomas D. *Culture Clash: Managing the Global High-Performance Team*. New York: Select Books, 2003.

MEET OUR EXPERT SOURCES

Snehal Antani

Snehal Antani serves as the senior vice president (SVP), IoT and Business Analytics, for Splunk. He joined Splunk in 2015 as chief technology officer (CTO) and helped drive the long-term vision and strategy for the company across IT, Security, Business Analytics, and the Internet of Things.

Prior to Splunk, he served as CIO of GE Capital's Distribution Finance business, as well as chief architect for GE Capital North America. In 2016, he was recognized as a Premier 100 award winner by Computerworld, for the digital transformation he drove while at GE, evolving IT from a back-office function to a core part of the value delivered to customers.

He began his career as a software engineer at IBM, where his work led to 12 patents spanning systems optimization, data processing, and large-scale transaction systems.

Snehal serves on the board of directors for NeverAgain.org, a nonprofit focused on applying technology to help people in danger. In 2017, he was recognized by the National Diversity Council as a Top 50 Multicultural Leader in Technology.

He holds a bachelor's degree in Computer Science from Purdue University, and a master's degree in Computer Science from Rensselaer Polytechnic Institute (RPI).

Ingrid-Helen Arnold

Ingrid-Helen Arnold was appointed president of SAP's Data Network by the executive board in April 2016 to build a new "data-as-a-service" business for SAP globally, capitalizing on her strong background in leading SAP's digital transformation as CIO and chief process officer. In building up this new growth engine for SAP, she works closely with the executive board.

From May 2014 to April 2016, Ingrid-Helen served as SAP's chief information officer (CIO) and chief process officer (CPO) for SAP SE as a member of the SAP Global Managing Board.

As an innovation leader with comprehensive experience in reinventing the way SAP runs its business, she drove SAP's continuous innovation journey in streamlining and rethinking internal processes and systems at SAP. As CIO and CPO, she was responsible for the adoption of SAP's solution portfolio internally and, as such, renovated key business processes to match SAP's cloud and platform strategy.

Ingrid-Helen worked in close alignment with the business and development to ensure that SAP software is user-centric and innovative. She also ensured that processes and systems are closely aligned in order to strengthen enablement and efficiency across SAP's lines of business.

Ingrid-Helen headed Business Innovation and Application Services from 2014 on, and led SAP cloud delivery from May 2014 to March 2015, operating powerful private (HANA Enterprise Cloud) and public managed cloud solutions for SAP customers. From 2012 to 2013, she headed Enterprise Analytics & Innovative Solutions at SAP, driving the internal adoption agenda of innovative solutions for SAP and leading several board programs.

Before that, she held various positions within SAP, including business controller for the Global Consulting Organization and COO for Global Controlling. She began her career in the finance department at Lafarge (Canada) before joining SAP in 1996.

She holds a master's degree in Business Studies from the University of Applied Sciences, Ludwigshafen.

Ramón Baez

Ramón Baez retired from Hewlett Packard Enterprise in October 2016, where he served as the senior vice president, customer evangelist, and customer advocate for Hewlett Packard Enterprise.

In this role, Ramón was championing customer centricity. Hewlett Packard Enterprise believes in relationships to drive the company's business. He advised customers and partners through frank peer-to-peer discussions on how to make and implement decisions about emerging digital technologies in their operations, products, and business models, which will have a positive impact in the marketplace.

His career spans 40 years at global Fortune 100 companies in the manufacturing, packaged goods, aerospace and defense industries, and products and services for the scientific community. Prior to HP, he held CIO positions at Kimberly Clark Corp., Thermo Fisher Scientific, Inc., and Honeywell's Automation and Control Solutions group. Ramón began his career at Northrop Grumman, where he spent 25 years and finished as CIO for its electronics systems sensor sector.

Ramón joined Hewlett Packard in 2012, as a SVP and global CIO, a role in which he oversaw the worldwide information technology (IT) strategy and all of the company's IT assets that supported Hewlett Packard employees and helped to drive strategic company priorities. This included global application development, the company's private cloud, IT security, data management, technology infrastructure, and telecommunication networks.

He also served as chairman of enterprise-wide Diversity & Inclusion Board, an entity that serves as the strategic governing body for global diversity and inclusion (D&I) at Hewlett Packard Enterprise. The board works closely with company leaders to champion the company's commitment to diversity and inclusion.

He continues to serve on boards of DocuSign, Hispanic IT Executive Council (HITEC) Foundation, and Trpz.com, as well as his most recent appointment to the board of directors for Kaiser Foundation Health Plan and Kaiser Foundation Hospitals. Additionally, Ramón is co-chair of HMG Strategy's Thought Leadership and CIO evangelist.

He holds a bachelor's degree in business administration from the University of La Verne in California.

Asheem Chandna

Asheem Chandna is a partner at Greylock Partners, where he is focused on helping entrepreneurs create and build the next generation of enterprise companies. His current companies include Aquantia, Avi Networks, Awake Security, Delphix, Innovium, Obsidian Security, Palo Alto Networks (PANW), Rubrik, and Skyhigh Networks.

Previous company boards and/or investments include AppDynamics (CSCO), Arista Networks (ANET), Aruba Networks (ARUN), CipherTrust (INTC), Imperva (IMPV), Net-Boost (INTC), PortAuthority Technologies (WBSN), Securent (CSCO), Sourcefire (CSCO), Sumo Logic, TechProcess (Ingenico), Xsigo Systems (ORCL), and Zenprise (CTXS).

Asheem joined Greylock in 2003 from Check Point Software (CHKP), where he was vice president of business development and product management. Asheem has been listed in the Forbes Midas List for the last six years.

Guy Chiarello

A renowned global leader in finance and commerce, Guy Chiarello has been at the forefront of technology innovation for more than 20 years. Guy has led product development and technology for three Fortune 500 companies, advised some of the biggest names in Silicon Valley and Wall Street, and has been recognized around the world as one of the most prominent technology leaders in the industry.

Guy is the president of First Data Corporation, a global leader in commerce technology. Since arriving at First Data in 2013, Guy's leadership has driven a transformation in First Data product development, bringing numerous solutions to market and acquiring several technology startups. Guy also played an integral role in guiding First Data to the largest US IPO in 2015 when the company raised $2.8 billion.

Prior to joining First Data, Guy was the CIO at JPMorgan Chase from 2007 to 2013, where he led global technology across all lines of business in more than 60 countries. Guy also served as the CIO of Morgan Stanley, where he began his career in financial services.

He is a graduate of The College of New Jersey (TCNJ), where he has received distinction for both athletic and professional achievements. Guy has been happily married to Denise for more than 30 years, and his greatest joy is spending time with his wife, children, and grandchildren.

Lee Congdon

Lee Congdon is senior vice president and CIO at Ellucian, the leading independent provider of higher education software, services, and analytics. He is responsible for Ellucian's information technology, including enabling the business through technology services, information technology strategy, delivering next generation solutions, process improvement, and advanced data and analytics.

Before joining Ellucian, Congdon was CIO at Red Hat. He led the implementation of many innovative open source and hybrid cloud solutions in a rapidly growing, dynamic environment. He enabled and supported the business through services such as customer engagement, knowledge management, technology innovation, and technology-enabled collaboration.

Prior to joining Red Hat, he was managing vice president, information technology, at Capital One, where he developed and delivered IT solutions for the firm's corporate functions and Global Financial Services group.

Earlier, Congdon was senior vice president, Strategic Initiatives, at Nasdaq, where he led the enterprise's efforts to identify, implement, and operate technology solutions for strategic global ventures. He holds B.S. and M.S. degrees in Computer Science from Purdue University and a master's of business administration (MBA) from Northwestern University.

Jason G. Cooper

Jason Cooper most recently served as vice president and chief analytics officer for Horizon Healthcare Services Inc., where he was responsible for enterprise-wide data analytics and informatics.

Jason is a 20-plus-year data and analytics professional and seasoned business executive specializing in analytics and informatics covering for-profit, nonprofit, and government domains, including leading teams at Blue Cross Blue Shield plans, Cigna and CVS Health, as well as experience with advanced NIOSH, NASA, DOD, and other software systems. He is a well-published author, experienced public speaker, and researcher.

Additionally, Jason is a member of the International Institute for Analytics, the Healthcare Information and Management Systems Society, the Healthcare Financial Management Association, and Sentrian's Advisory Board. He holds master's degrees in Computer Science and Biomedical Engineering.

Mignona Cote

Mignona Cote is a proven problem solver with a unique ability to drive solutions for information security vulnerabilities across a variety of industries. With over two decades of experience in information security, risk management, compliance, and auditing, she has transformed technical operations and built strategic solutions for Fortune 50 companies like Aetna, Bank of America, PepsiCo, and Verizon.

Today, Mignona serves as the CISO for two Aetna subsidiaries, PayFlex (healthcare financial products) and Phoenix Data Services (IT hosting for subsidiaries). She leads Aetna's Policy Management program, where she's responsible for enterprise risk management governance, subsidiary security governance, security education, policy management, regulatory response, and evangelizing security among Aetna's top corporate customers.

Kirsten Davies

Kirsten Davies is the newly appointed CSO/CISO for Barclays Africa, with the mandate to build and run the end-to-end Security Program for operations on the continent, including cybersecurity, information security, fraud, forensics and investigations, physical security, and resiliency. With the decoupling of Barclays Africa from the global entity, her program will be the first of its kind on the continent for this multicountry, stand alone, financial services entity.

Kirsten brings a unique blend of leadership savvy, business acumen, security practitioner experience, and technology innovation to her role. She is a proven leader in guiding large-scale global transformation programs focused on cyber and information security, technology, business process, and organizational transformation. Africa is the fourth continent on which she has lived and worked, providing truly unique perspectives on and successful delivery within global cultures, customs, practices, and business operations.

Prior to her role with Barclays Africa, she served in two capacities at Hewlett Packard Enterprise, both as vice president & deputy chief information security officer (CISO) and as vice president for Enterprise Security Strategy, leading one of four worldwide, strategic, customer-facing transformation initiatives for HPE.

In her tenure with HPE, she spearheaded the first-ever Cyber Security Master Agreement with the German Workers Council, an agreement that is now replicated across HPE's 20+ Works Councils in EMEA. She also led the strategic redesign and reorganization of the Cyber Security Program, and infused new breadth and depth into the "Next Generation Security" areas of Insider Risk and the IT/OT convergence.

Prior to HPE, Kirsten served as the CISO for Global InfoSec Strategy and Governance at Siemens. There she led the multiyear global InfoSec Transformation Program to refine the capabilities, policies, processes, culture, and technical portfolio of information security. Before that, she served in senior leadership positions at Booz Allen Hamilton and Deloitte Australia.

Kirsten holds a BA in International Politics & Government from University of Puget Sound in Tacoma, Washington, has studied International Law at Johns Hopkins

University and has an advanced certificate in Change Management from the PROSCI Institute. She speaks near-fluent German, is a wine enthusiast, and has a history of rescuing adorable dogs.

Dana Deasy

Dana Deasy is managing director and global CIO of JPMorgan Chase, responsible for the firm's technology systems and infrastructure across all of the firm's businesses globally. Dana manages a budget of more than $9 billion and over 40,000 technologists supporting JPMorgan Chase's Retail, Wholesale, and Asset Management businesses. He has more than 30 years of experience leading and delivering large-scale IT strategies and projects.

Prior to joining JPMorgan Chase in December 2013, Dana was CIO and group vice president at BP, where he was responsible for global information technology, procurement, and global real estate.

Earlier in his career, Dana served as CIO for General Motors North America, Tyco International, and Siemens Americas. He also held several senior leadership positions at Rockwell Space Systems Division, including as director of information management for Rockwell's space shuttle program.

Dana was inducted into the CIO Hall of Fame in 2012 and the International Association of Outsourcing Professionals Hall of Fame in 2013. He was also named Transformational CIO in 2017.

Dana serves as an executive in residence and member of the advisory committee for the Executive Master of Science in Technology Management Program at Columbia University. He holds an undergraduate degree from the University of Southern California and an MBA from National University.

Scott Fenton

Scott Fenton is the principal owner of Scott Fenton Consulting, LLC (www.sfentonconsulting.com). He has over 25 years of experience in IT. Scott has expertise in strategic planning and ensuring companies obtain the maximum return from their IT investment.

He previously held executive level positions at Intel, Wind River Systems, Peregrine Systems, HP, Fujitsu, and Tektronix. Fenton's consulting practice offers services related to interim CIO, cloud strategies, IT assessment, sales and marketing optimization, and strategic planning.

Fenton holds a master's degree in information management from the University of Oregon.

Scott has received repeated recognition and awards for technology innovation from CIO Magazine, Oracle, and InformationWeek.

John Foley

John's personal path to high performance began as a child, when he stood alongside his father at an air show featuring the Blue Angels. From that moment, John knew deep in his heart that someday he'd be carving up the skies as a member of the Blues. Eventually, he lived that dream, but getting there wasn't easy.

In fact, John's journey from an awestruck child at an air show to the cockpit of the Blue Angels' F/A-18 Hornet is a study in persistence, hard work, and the will to overcome obstacles and setbacks. Those ideals fit within three over-riding traits that mark John's presentations: (1) a contagious attitude of thankfulness that he calls Glad To Be Here®; (2) an energizing delivery that inspires High Performance and service to others; (3) a practical model for living out his message that works in other organizations as well as it works for the Blue Angels.

John graduated from the US Naval Academy with a degree in mechanical engineering. He also was a defensive back for the Midshipmen, playing in two bowl games and helping Navy to one of the best four-year records in its football history.

As a pilot, John was a "Top Ten Carrier Pilot" six times before becoming a Marine instructor pilot and a Blue Angel. He holds master's degrees in business management, from the Stanford Graduate School of Business (as a Sloan Fellow); in international policy studies, from Stanford University; and in strategic studies, from the Naval War College. He makes his home in Sun Valley, Idaho.

Pat Gelsinger

Pat Gelsinger has served as CEO of VMware since September 2012, almost doubling the size of the company during his tenure.

He brings with him more than 35 years of technology and leadership experience. Before joining VMware, Gelsinger led EMC's Information Infrastructure Products business as president and COO. Prior to EMC, Gelsinger spent 30 years at Intel, where he was the company's first CTO and drove the creation of key industry technologies, including USB and Wifi, led Intel to be the dominant supplier of microprocessors, and was the architect of the original 80486 processor.

Gelsinger earned an associate's degree from Lincoln Technical Institute in 1979, a bachelor's degree from Santa Clara University in 1983 (magna cum laude) and a master's degree from Stanford University in 1985, all in electrical engineering.

In 2008, he was named a Fellow of the IEEE and was awarded an Honorary Doctorate of Letters from William Jessup University. He holds seven patents in the areas of VLSI design, computer architecture, and communications. He is a recognized values-based leader, receiving the Gordon University Faithful Leader award, and a sought-after speaker on technology, culture, business, and faith-based topics globally.

Clark Golestani

Clark Golestani brings over 30 years of experience in technology and life sciences, and is currently the president, Emerging Businesses and Global CIO, for Merck (merck.com), with responsibilities for Merck's portfolio of digital health services and solutions companies inclusive of the company's venture and equity funds—each of which extensively leverages opportunities across the digital health ecosystem. Further, Clark leads the company's global IT organization, strategy, and execution worldwide, spanning all lines of business inclusive of Human Health and Animal Health BioPharma.

Additionally, Clark serves on the board of directors for Seal Software (seal-software.com), the leading provider of contract discovery, extraction, and analytics capabilities, the board of directors for UMUC Ventures (umucventures.org), a not-for-profit organization focused on advancing adult education, as well as the advisory councils for multiple venture funds and technology product and services companies.

Prior to joining Merck, Clark was responsible for establishing and managing strategic client relationships from business development through consulting services engagement delivery as a principal with Oracle Corporation's consulting practice.

Previously, Clark served on the board of directors for Liaison Technologies (liaison.com), and NPower (npower.org).

Clark has a degree in Management Science from Massachusetts Institute of Technology Sloan School of Management, and is a cofounder of Cross Road Technologies, Inc., in Cambridge, Massachusetts.

Kevin A. Haskew

Kevin A. Haskew joined ON Semiconductor in 2011 and is senior vice president and CIO responsible for the company's worldwide IT systems. He has built an effective and responsive global IT organization of approximately 750 associates.

He is a recognized technology leader known for implementing distinctive IT strategies and achieving aggressive business and corporate goals by offering an exceptional blend of executive insight, IT solutions development, and strategy-based team building.

While at ON Semiconductor, he has focused on forming a sound IT foundation that delivers results by aligning technology initiatives to corporate strategies with substantial improvement to standardization, service delivery and business systems improvements. He oversees the company's infrastructure, corporate application, manufacturing applications, technology security, compliance program, and global integration efforts through attained mergers and acquisitions.

In addition to key business process improvements garnering both respect and visibility from his stakeholders, his astute business sense has been instrumental in providing ON Semiconductor with IT capabilities critical to the success of manufacturing systems and company performance while providing cost containment even in the most tedious financial times.

Prior to ON Semiconductor, he was vice president and CIO for Biomet LLC ($3.8 billion), where he was a member of the executive team that transformed Biomet in preparation for an initial public offering (IPO). Prior to Biomet, he was vice president and CIO at ArivnMeritor Inc. ($10 billion).

Additionally, he was director of IT at Lear Corporation, and prior to that, he held various IT positions within Allied Signal, Inc. (which later became Honeywell), living overseas.

Kevin holds a master of science (MSc) in IT and change management and an MBA. He also maintains memberships in the British Computer Society as a fellow, Chartered IT Professional (CITP), and Chartered Engineer (CEng) and sits on the board of managers for Desert Schools Financial Services LLC.

Patty Hatter

Patty Hatter is SVP and general manager of the McAfee Professional Services organization. She recently transitioned from the role of general manager for Intel Security and Software IT & McAfee CIO, and prior to that was the SVP of Operations and CIO at McAfee. She has overall responsibility for leading the professional services organization and expanding McAfee's consulting, managed services, deployment, and training services.

Patty is a visionary IT leader who designs and deploys transformational strategies for global companies that accelerate revenue, market position, and profitability.

At McAfee and Intel Security, she transformed IT from an order-taking organization to a proactive, integrated, thought-leading function that put expanded and enhanced capabilities and infrastructure into the business and product teams.

Patty earned bachelor's and master's degrees in mechanical engineering from Carnegie Mellon University. She currently holds multiple advisory board positions, and is also a board member for the Silicon Valley Education Foundation.

Shawn Henry

As president of CrowdStrike Services, Shawn leads a world-class team of cybersecurity professionals who aggressively and effectively investigate and mitigate targeted attacks on computer networks. Under his leadership, CrowdStrike has been engaged in significant proactive and incident response operations across every major commercial sector, protecting organizations' sensitive data and networks.

Shawn retired as FBI executive assistant director (EAD) in 2012, overseeing half of the FBI's investigative operations, including all FBI criminal and cyber investigations worldwide, international operations, and the FBI's critical incident response to major investigations and disasters. During his 24-year career, he held a wide range of operational and leadership roles in four FBI field offices and FBI headquarters.

Serving in multiple positions relating to cyber intrusions since 1999, Shawn was the Bureau's outspoken top agent on cybersecurity issues, boosting the FBI's cyber investigative capabilities. In addition to his last position as EAD, he served as both deputy assistant director and assistant director of the cyber division at FBI headquarters; supervisor of the

FBI Cyber Crime Squad in Baltimore; and chief of the Computer Investigations' Unit within the National Infrastructure Protection Center (NIPC).

During his tenure, Shawn oversaw major computer crime and cyber investigations spanning the globe, from denial-of-service attacks, to major bank and corporate breaches, to nation-state sponsored intrusions. Shawn led the establishment of the National Cyber Investigative Joint Task Force (NCIJTF), a multiagency center led by the FBI, and forged partnerships domestically and internationally within governments and the private sector.

He was an original member of, and key contributor to, the National Cyber Study Group, under the direction of the Office of the Director of National Intelligence. This organization developed the Comprehensive National Cybersecurity Initiative (CNCI), the US government's national strategy to mitigate threats and secure cyberspace. Early in his cyber career, Shawn served on the US delegation to the G8 as a member of the High-Tech Crimes Subgroup.

Shawn has been a keynote speaker in some of the largest cyber conferences in venues around the world, and he has been interviewed extensively on radio and quoted in numerous print and online publications, including the Associated Press, *Bloomberg Businessweek, Forbes, The New York Times, Reuters, USA Today, The Wall Street Journal,* and *The Washington Post,* among many others. Shawn has been a regular contributor on every major domestic television network, and is currently a network news analyst exclusively for NBC.

Shawn's professional achievements have been recognized throughout his career. In 2009, he received the Presidential Rank Award for Meritorious Executive for his leadership in enhancing the FBI's cyber capabilities, and SC Magazine recognized him as one of the top industry pioneers who shaped the information security industry. In 2010, he was named one of the most influential people in security by *Security Magazine*; received the Federal 100 Award as a government leader who played a pivotal role in the federal government IT community; and was selected as cybercrime fighter of the year by McAfee Inc.

Shawn has lectured on a variety of policy and strategy topics at major universities, including American, George Washington, Georgetown, Harvard, and Johns Hopkins. He served previously as a director on the board of SecureBuy, and currently sits on the cyber advisory board to the governor of New York and the Global Cyber Alliance. Shawn is on the faculty and a Board Leadership Fellow at the National Association of Corporate Directors (NACD), where he speaks to corporate boards and directors about complex cybersecurity issues.

Shawn earned a bachelor of business administration from Hofstra University and a master of science in Criminal Justice Administration from Virginia Commonwealth University. He is a graduate of the Homeland Security Executive Leadership Program of the Naval Postgraduate School's Center for Homeland Defense and Security.

Zack Hicks

Zack Hicks is CEO and president of Toyota Connected and SVP and CIO of Toyota Motor North America (TMNA).

As CEO and president of Toyota Connected, he leads global efforts, reinventing the way we think about mobility, developing solutions that excite our customers and anticipate their needs, by utilizing advanced technology and through the art and science of predictive intelligence.

In his CIO role, Hicks drives the strategy, development, and operations of all systems and technology for Toyota's North American operations. Hicks is focused on aligning the efforts of business operations, strategic planning, and technology to drive business innovation and efficiency.

Hicks is also a member of the North American Management Committee, board member of the Toyota Foundation and Toyota Insurance Management, and chairs the Executive Steering Committee.

Hicks received Pepperdine's Distinguished Alumnus Award at the Graziadio School of Business and Management. He has been named by *CIO Magazine* as Ones to Watch, awarded multiple recognitions by *CIO Magazine*'s CIO 100 list, was one of the top 10 Global Breakaway CIO Leaders by Evanta/CIO Leadership Network, and was named a Premier 100 IT Leader by *Computerworld*.

He is an active member of the CIO Strategy Exchange, Corporate Executive Board for Chief Information Officers, CIO Leadership Network, and Trevor Project Board of Directors and Research Board.

Adriana Karaboutis

Adriana (Andi) Karaboutis is recognized for her success as an innovative technology executive and a business leader across a range of sectors that include industry giants, such as National Grid, Biogen, Dell, Ford, and General Motors. She was recently appointed chief information and digital officer for National Grid. As a member of the company's executive committee, she will be responsible for the development of an enterprise-wide digital strategy, delivery of information systems and services, digital security and risk, as well as overall security.

Previously, as executive vice president for Technology, Business Solutions, and Corporate Affairs at Biogen, she had a broad set of responsibilities that included information technology, digital health and data sciences, and corporate affairs.

Prior to joining Biogen in 2014, she was vice president and global CIO of Dell, Inc. and previously spent more than 20 years at General Motors and Ford Motor Company in various international leadership positions, including computer-integrated manufacturing, supply chain operations, and information technology.

She is also currently on the board of directors of Perrigo Company plc, Advance Auto Parts, and Blue Cross Blue Shield of Massachusetts. Additionally, she has served as president of the Michigan Council of Women in Technology (MCWT); a board member of the Manufacturing Executive Leadership; and on the Babson College advisory board for the Center for Women's Entrepreneurial Leadership (CWEL).

She received a bachelor of science degree in Computer Science from Wayne State University in Detroit, and was a Merit Scholar. Additional education includes graduate courses at Wayne State and completing the accelerated Marketing Strategy Program at Fuqua School of Business (Duke University).

Timothy Kasbe

Timothy Kasbe is the chief information and digital officer for the Warehouse Group. He is a strategic leader and innovator with digital and business acumen gathered from working globally and leading teams across diverse cultures.

He has led large-scale digital transformations in retail, high-tech, and finance sectors, having impacted companies including Westpac, IBM, Reliance Industries Ltd., Sears Holdings, and Gloria Jeans.

Prior to joining the Warehouse Group, Timothy was COO of Gloria Jeans Company. He led all aspects of the business including technology, supply chain, sourcing, and operations. Timothy lives the value that we exist to delight our customers with the beauty of technology and community experiences.

In his role at The Warehouse Group, Timothy will lead transformation of the Group into a customer-focused digital retail company. He serves on the board of Tamr Inc., a Cambridge, Massachusetts–based data visualization and machine learning company. He is a distinguished visiting scholar at Stanford University, in California.

Ralph Loura

Ralph Loura is CTO at Rodan + Fields, where he defines and advances the company's technology strategy and infrastructure to support significant growth momentum and deliver user-friendly digital experiences. Ralph leads the Rodan + Fields innovative technology platform providing customers with a high-touch, high-tech experience.

Most recently, he served as vice president and CIO for Enterprise and Global Sales Operations at Hewlett-Packard (HP), where he led business intelligence, operations, and customer relationship management teams in identifying market opportunities and implementing go-to-market strategies.

Prior to HP, Ralph spent several years as SVP and CIO of The Clorox Company, implementing technology initiatives to support the company's overall growth.

Earlier in his career, he held technology leadership roles at Cisco, Symbol, and AT&T Bell Laboratories, among others. He has received numerous industry awards, including *Computerworld's* 2012 Premier 100 IT Leaders and *Consumer Goods*

Technology's 2013 CIO of the Year, and currently serves on the board of Big Brothers Big Sisters of the Bay Area.

Ralph holds an M.S. in Computer Science from Northwestern University and a B.S. in Mathematics and Computer Science from St. Joseph's College.

Shamim Mohammad

Shamim Mohammad joined CarMax in 2012 as vice president of application development and IT planning. In 2014, he was promoted to SVP and CIO.

In his current role, Shamim is responsible for all strategic use of technology throughout the company, including managing applications, information security, architecture, infrastructure, operations, and business planning. As a member of the senior executive team, he leads business transformations enabled by technology.

Prior to working at CarMax, Shamim was the vice president of information technology for BJ's Wholesale Club, and worked in various senior IT leadership positions for Blockbuster and TravelClick.

Shamim received his bachelor of science degree in Computer Science with concentration in Accounting from Angelo State University, and his MBA from Kellogg School of Management at Northwestern University. He is a registered Certified Public Accountant (CPA) in the state of Illinois.

He serves on the board of directors for Richmond Technology Council, CodeRVA, and the YMCA of Greater Richmond.

Tom Peck

Tom Peck is SVP and CIO at AECOM (NYSE: ACM), a $20-billion global provider of professional technical and management support services. AECOM's 100,000 employees— including architects, engineers, designers, planners, scientists, and management and construction services professionals— design, build, finance, and operate infrastructure assets for public and private sector clients in more than 150 countries around the world.

In this position, Tom is responsible for the support and deployment of technology to fuel growth and efficiencies for both employees and clients. His goal is to make technology a competitive advantage at AECOM. He works side-by-side with operational leadership to ensure effective infrastructure and support as well as strategic alignment of the company's technological platforms and investments. From 2013 through 2017, he also led the company's global procurement and travel departments where strategic sourcing, spend controls, and lean supply chains power AECOM's operating efficiencies. He is a member of the CEO's executive committee.

Before joining AECOM, he was the CIO at Levi Strauss & Company, which operated brands across 60,000 sales locations in 110 countries. Prior to Levi Strauss, he was the CIO at MGM Mirage, where he led an IT transformation and an expansion of mega-resorts.

Prior to these roles, he had an accomplished career at General Electric, culminating in the CIO role for NBC Universal's global entertainment business unit. He also served as the global quality leader for NBC Universal.

Tom has a history of innovation, is a frequently requested keynote speaker, and was inducted into the CIO Hall of Fame in 2015.

A former financial management officer in the US Marine Corps, Tom had a distinguished career inclusive of both operational and headquarter assignments culminating in an enterprise-wide program management role on the Marine Corps CFO's team at the Pentagon.

Mark Polansky

Mark Polansky is a senior partner in Korn Ferry's Information Technology Officers Center of Expertise, a practice he headed up for more than 12 years. He has been based in the firm's New York office since joining Korn Ferry in 1997.

Mark has extensively recruited CIOs, CTOs, CISOs, and other senior IT leadership talent across varied vertical sectors, including industrial, consumer, life sciences, technology, and higher education. His consultative approach to search encompasses talent strategy, organization design, search process, and diversity recruiting.

Mark's clients range from the Fortune 100 (Intel, McKesson, General Electric, Hewlett-Packard, Boeing) to smaller high-growth companies (Under Armour, Regeneron, DirecTV) as well as prestigious public and private universities (Harvard, Yale, Rockefeller, Florida).

Before entering the search field, Mark developed and marketed information systems within the Financial Services and Higher Education sectors. He currently serves on the Advisory Committee for Columbia University's executive graduate program in Information Technology Management.

Mark frequently addresses conferences and writes on information technology and career management topics. He is the creator of the "Executive Career Counsel" column in *CIO Magazine* and on CIO.com. He is a member of the Society for Information Management (SIM) and has served as president of the New York chapter. He sits on the advisory boards of The Information Technology Senior Management Forum (ITSMF), the national organization dedicated to fostering executive talent among African-American IT professionals, and HITEC, the Hispanic Information Technology Executive Council.

Mark holds a master's degree in Computer Science from Pratt Institute and a bachelor's degree in Mathematics and Electrical Engineering from Union College.

John Rossman

John Rossman is the author of *The Amazon Way* book series, and a managing director at Alvarez and Marsal, a global professional services firm. Mr. Rossman is an expert at crafting and implementing innovative and digital business models and capabilities, including the Internet of Things. He is a sought-after speaker on creating a culture of operational excellence and innovation.

Prior to joining A&M, John was an executive at Amazon.com, where he launched the marketplace business, and ran the enterprise services business. Mr. Rossman's blog is www.the-amazon-way.com and can be reached at johnerossman@gmail.com.

Bernadette Rotolo

Bernadette Rotolo is a global Fortune 500 executive with nearly 20 years of experience in technology at Accenture, Adecco, and Warner Music Group. Bernadette started her career in consulting at Accenture, a top IT consulting company assisting clients in communications and high-tech industries with their most difficult technology and business challenges. After spending 12 years at Accenture, Bernadette came to Adecco Group in 2009, a Swiss-based Global Fortune 500 Company that is the largest staffing company in the world. Bernadette was a group vice president and Global Head of Solutions Delivery Management focused on digital offerings.

For the last four years at Adecco, Bernadette led a CEO-championed global digital transformation program partnering with business leaders to increase brand visibility by launching web, mobile, and social media strategies that increased overall global millennial engagement. In early 2017, Bernadette joined Warner Music Group as its SVP, Head of Global Systems, reporting to the global CIO to design and lead another global transformational program.

Bernadette holds her master's degree from New York University and her bachelor's from the State University of New York at Binghamton. She also earned a leadership certification from the prestigious IMD Business School in Lausanne, Switzerland. Bernadette is also a member of the Society for Information Management (SIM) Metro New York chapter and an active member of the HMG network.

William Ruh

Mr. William Ruh is the CEO for GE Digital as well as the SVP and chief digital officer (CDO) for GE. GE Digital, a $6 billion business of General Electric, provides premier digital software solutions and services for the industrial world. GE Digital supports customers globally with a broad range of Industrial Internet applications, from asset performance management, operations optimization, and brilliant manufacturing to platform-as-a-service, cloud, and cybersecurity. As the CDO, Bill is responsible for global IT as well as creating GE's Digital Thread, a next-generation system for streamlining design, manufacturing, and support processes.

Bill joined GE in 2011 to establish its Industrial Internet strategy and to lead the convergence of the physical and digital worlds within GE globally. In this role, he focused on building out advanced software and analytics capabilities, as well as driving the global strategy, operations, and portfolio of software services across all of GE's businesses. During his tenure, he led the charge to develop the first cloud-based platform for the industrial world. He also played an instrumental role in establishing the Industrial Internet Consortium by bringing together government, academia, and industry leaders for setting the standards, best practices, and processes for the Industrial Internet.

Prior to this, he was vice president at Cisco where he held global responsibility for developing advanced services and solutions. A 30-year veteran of the software and internet industries, he has held executive management positions at Software AG, Inc., The Advisory Board, The MITRE Corporation, and Concept 5 Technologies. Bill is an accomplished author and a frequent speaker on such topics as emerging business models, cloud computing, analytics, mobile computing, agile development, large scale distributed systems, and M2M communications. He earned a bachelor's and master's degrees in Computer Science from California State University, Fullerton.

Kevin Sealy

Kevin Sealy is a senior client partner in Korn Ferry's London office, where he leads the firm's CIO & IT Officers Practice in EMEA. Kevin has over 30 years of experience in the information technology industry, working cross-sector, both as practitioner, consultant, and search professional.

He specializes in appointing CIOs and senior information technology professionals across all industry sectors, and has worked with many leading global companies on high-profile CIO-related placements. In his previous position as a partner in IBM Business Consulting (formerly PwC Consulting), he led the delivery of technology strategy and systems integration work with blue-chip clients internationally.

Kevin began his career as a graduate trainee in the IT department at Mobil Oil Company Ltd. He holds a master's degree in physics from Oxford University, and is a Member of the Institute of Consulting (MIC) and a Certified Management Consultant (CMC).

Naresh Shanker

Naresh Shanker is CIO of HP, where he leads the company's global IT strategy and operations. His responsibilities include application development, data management, technology infrastructure, data center operations, and telecommunication networks to enable the company's digital transformation.

Naresh's experience within the IT industry spans more than 25 years and incorporates supporting a wide range of industries including high-tech manufacturing, semiconductor, medical, life sciences, and software businesses. He has also led and executed large scale software implementations, outsourcing, mergers and acquisitions, and has established offshore IT and manufacturing operations.

Most recently, Naresh was vice president and CIO of the HP Printing & Personal Systems Group, which included IT responsibilities for engineering and globally deploying

Supply Chain, HP.com, Marketing, Communications, and Big Data/Analytics solutions.

Prior to joining HP's Printing and Personal Systems Group, Naresh was vice president and CIO for Palm, Inc., where he had management responsibility for the IT strategy and all of the company's IT assets. In addition, Naresh was instrumental in launching the webOS products and platform to enable the App Catalog, mobile commerce and cloud services, as well as webOS.com e-commerce operations.

Prior to joining Palm, Naresh was responsible for managing Agilent Technologies' enterprise business solutions organization, delivering customer-facing and supply-chain solutions worldwide.

Naresh began his career at HP in the Medical Products Group, where he led the GBU's IT team that was responsible for delivering applications and infrastructure solutions and services worldwide, as well as integrating acquisitions into the business. He also was instrumental in establishing a joint venture manufacturing center in Qingdao, China.

Naresh earned a bachelor's of science degree in Computer Science and a MBA from the Illinois Institute of Technology in Chicago.

Raj Singh

As CIO at FordDirect, a joint venture between Ford Motor Company and Ford franchise dealers, Rajoversees an IT department of professionals and 50+ technology providers and vendors at FordDirect.

He is responsible for application development and management, product developments, technology compliance, infrastructure management, data security, and technical innovation. Raj leads an IT annual budget of more than $110 million. Some of his key responsibilities and systems involve Digital Strategy, Lead Management, Inventory Management, Incentives, Online-Offline Integration, Dealer Digital Footprint Management, ERP Systems, Data Platform and Analytics.

Prior to joining FordDirect, he was the executive director of IT at Consumers Energy, where he led a portfolio of more than 290+ applications and over 400+ global technology resources, including vendor partner resources. He led a team of over 175+ employees and over 200+ consultants. Singh was also responsible for building a flexible, nimble delivery model, including business intelligence and IT transformation.

Raj holds a bachelor's degree in computer science, an MBA from Michigan State University, and a master of program management from George Washington University. He is also project management professional-certified (PMP), SAP COE-certified, and a Six Sigma Green Belt.

Jim Swanson

James Swanson is CIO for Monsanto, a leading global agricultural company. Jim leads the 1,200-plus employees across the global Information Technology function. Jim is responsible for the IT vision to transform Monsanto into an information-based company and deliver IT capabilities across all Monsanto's global business. Jim is a member of Monsanto's Corporate Strategy Leadership Team, Global Business Operations Leadership Team, and Executive Team Operations Council.

Jim's previous work experience included executive and scientific roles at Merck, Johnson & Johnson, and SmithKline Beecham.

Joe Topinka

Joe Topinka is a dynamic author, speaker, coach, and career CIO. He has over 35 years of success leading IT organizations and driving meaningful business results in the process. His game-changing IT Business Partner Program bridges the chasm between business stakeholders and IT organizations. Joe is a frequent keynote speaker on this subject where he shares his vision at industry events across the country. His bestselling book, *IT Business Partnerships: A Field Guide,* was named a top-10 read for CIOs in 2014.

Joe is board chair of the Business Relationship Management Institute, an organization dedicated to the advancement of BRM capability worldwide. Joe is also the founder of CIO Mentor, LLC, a consultancy focused on helping companies leverage the power of technology to achieve profit-driven business results. He is a two-time CIO of the Year award winner in both Charlotte (2015) and Minneapolis (2013). He is currently VP-CIO of SnapAV and is responsible for an innovative award-winning IoT-based platform call OvrC (pronounced oversee) that is changing the way home and SMB integrator serve their customers.

Michael Wilson

Michael has 25 years' experience in the Information Technology and Risk management; for the last eight years with McKesson Corporation where he is responsible for the security, business continuity and IT risk management functions for the Corporation. McKesson is a Fortune 5 ranked, $181-billion, 85,000-employee company; providing global Health Distribution and Technology Services for its' customers. Since joining McKesson, Mike has worked to build a team to respond to an ever-increasing complex global regulatory environment and thwarting criminal attempts to compromise McKesson's systems.

Mike has experience across several geographies and industries, including financial services, insurance, healthcare, utilities, and consumer products. He is regarded as a national thought leader and was voted by his peers as a 2011 CISO of the year for his industry contributions toward cybersecurity awareness. He is a member of several advisory boards and industry organizations.

Prior to joining McKesson, he worked for many years for a global professional services organization and held various IT positions in the AsiaPac region working in the Financial Services industry for a Bank and General Insurance company. Michael holds a bachelor's degree in Business Commerce and Administration from Victoria University, New Zealand, MBA from Massachusetts Institute of Technology (MIT) Sloan School of Management, and Graduate Certificate, Health Management (MIT).

Deanna L. Wise

Deanna Wise, executive vice president and CIO for Dignity Health, has been in the information technology field since 1989 and health care IT since 1994. She has led the Information Technology due-diligence for numerous individual hospitals and hospital network acquisitions.

She has a deep understanding and vast experience in the development and maintenance of industry-leading clinical IT systems. She plays a key role in Dignity Health's Horizon 2020 goals of competing in a reformed health care system that includes industry-leading clinical information technology.

Deanna oversees the full implementation of Dignity Health's EHR software standard, which brings state-of-the-art electronic records to each Dignity Health facility. She also oversees meeting meaningful use standards and is building electronic bridges between facilities and community physicians. She is committed to developing customer focus, efficiency, and excellence of the IT function in supporting a positive patient experience.

Before joining Dignity Health, Deanna was the CIO for Vanguard Health Systems. Prior to this role, she was the CIO for Vanguard's Abrazo Health Care Phoenix market.

During her time with Vanguard Health Systems, she succeeded in developing a staff that steadily improved customer satisfaction while improving the retention and satisfaction scores of the IT team. The IT team excelled in delivery of its most important technology projects, including the McKesson Clinical suite of products with a standard deployment as well as numerous Athena Ambulatory EMR deployments. This was accomplished while meeting project timelines and budget goals, preparing Vanguard to successfully meet the new ARRA guidelines.

Prior to joining Abrazo Health Care, she served as the CIO for the Maricopa County Health District and successfully in-sourced the entire Information Technology organization from McKesson and reduced the overall budget while improving customer satisfaction scores.

She also served as the director of Applications for the 12 Indiana regional facilities of Ascension Health. Wise earned a computer science degree and is a project management professional (PMP) certified member of the Professional Management Institute (PMI). She has successfully introduced the PMI methodology for large hospital technology projects at Ascension Health and Maricopa County as well as Vanguard Health Systems for system implementations of McKesson, Cerner, PeopleSoft, Lawson, Siemens, and other major companies.

ABOUT THE AUTHOR

Hunter Muller is a charismatic industry pioneer and energized leader. As a global C-suite advisor and coach for the past 30 years, Hunter's insights and concepts have been successfully applied by Fortune 1000 organizations worldwide to accelerate executive performance, propel the career ascent of C-level executives, and to drive transformative innovation and create new waves of value.

Hunter is the president and CEO of HMG Strategy, LLC, the world's largest independent provider of thought leadership and networking events for CIOs, CISOs, and senior technology executives. HMG Strategy offers a unique, integrated model for enabling C-level executives and their companies to draw upon courageous leadership to reimagine, reinvent, and rebuild their business models to attain the genius-level thinking required in today's dynamic business landscape to achieve competitive advantage in the digital frontier.

Complementing HMG Strategy's world-renowned events is HMG Ventures, a venture capital unit aimed at helping connect CIOs with the most innovative technology companies and entrepreneurs from Silicon Valley to Tel Aviv.

HMG Strategy also partners with the world's leading venture capital firms, including Amplify Partners, Greylock Partners, Lightspeed Venture Partners, and Sequoia Capital. HMG Strategy connects CEOs, CIOs, and other C-level executives with start-up technologies backed by the leading VCs to identify opportunities for innovation and business transformation.

In addition, HMG Strategy has partnered with Newtek to provide top-tier cloud, infrastructure, and cutting-edge technology services aimed at enabling clients to achieve digital dominance.

Meanwhile, HMG Strategy's award-winning Research Team advises the world's top technology companies such as Workday, ServiceNow, and Adobe on branding, messaging, and voice of the customer strategies.

Hunter has authored several books that draw from his extensive business experience: *The Transformational CIO* (Wiley, March 2011), *On Top of the Cloud* (Wiley, January 2012), *Leading the Epic Revolution* (Wiley, September 2013), *The Big Shift in IT Leadership* (Wiley, July 2015), and *The CEO of Technology* (Wiley, December 2017). As a result of Hunter's unique knowledge of the thoughts, challenges, and opportunities of the IT leader, he is asked to speak at a variety of venues and conferences.

Hunter and HMG Strategy support a variety of charitable causes, including the American Cancer Society, Save the Children, and Year Up, and Hunter personally gives back his time and ideas as a member of the executive committee of the Fairfield/Westchester Chapter of the Society for Information Management.

An intrepid international athlete, Hunter is an ardent yacht racer, a world-class skier, an extreme fitness guru, and a former Mr. America. Hunter is passionate about ocean and mountain adventures and spending time with his sons, Chase and Brice.

Hunter's adventurous spirit has prompted him to launch a new business venture—HMG Adventures—designed to enable top-tier executives to connect with their inner selves and discover fresh approaches to expand their leadership capabilities on the path to genius.

Upcoming events offered through HMG Adventures include world-class skiing in Snowbird, a heart-pounding Formula 1 Grand Prix race in Monaco, and action-packed high-mountain heli-skiing.

Hunter earned a bachelor of science degree at the Babson College School of Management.

ABOUT HMG STRATEGY

HMG Strategy is the world's largest independent and most trusted provider of executive networking events and thought leadership to support the 360-degree needs of technology leaders. Our regional CIO and CISO Executive Leadership Series, newsletters, authored books, and online **Resource Center** deliver proprietary research on leadership, innovation, transformation, and career ascent.

The HMG Strategy global network consists of over 300,000 senior IT executives, industry experts, and world-class thought leaders. Additionally, our partnerships with the world's leading search firms provide vital insights into the evolving roles of the CIO and CISO.

The HMG Strategy **CIO Executive Leadership Series** offers a completely unique experience for IT executives to gain the latest insights and best practices for driving increased business value through the use of IT and build invaluable relationships with peers and industry experts. The HMG Strategy **CISO Executive Leadership Series** is designed to provide information security leaders with the insights and best practices they need to tackle the most pressing cybersecurity challenges facing the enterprise today and going forward.

221

For more information about joining the strongest executive leadership network and exploring our unique online Resource Center, please visit www.hmgstrategy.com.

About CELA

The CIO and CISO Executive Leadership Alliance (CELA) is the trusted, independent source for addressing your top challenges. HMG Strategy curates, analyzes, and provides CELA members with the insights, trends, and emerging developments drawn from the industry's leading CIOs and CISOs that can help you to succeed in today's ever-changing business environments. HMG's expert services will distill actionable information you can apply in your role as a leader.

Key benefits of the CELA program to members:

- The invitation-only CELA program connects world-class IT leaders with one another to share their top challenges and proven methods for addressing them

- Becoming C-suite and boardroom ready

- Paving the right digital roadmap that maps with your company's top business goals

- Mastering IT cost controls, creating business model innovation, and delivering value to the enterprise

- Making a successful transition from legacy systems to a digital platform that creates greater agility and new business opportunities

- Strengthening your personal brand and powering your career ascent

- Cultivating proactive security strategies to safeguard the enterprise

INDEX